T0133216

Dieter Beule

Thermodynamic Properties and
Stochastic Kinetics of Ionization and Recombination
in Partially Ionized Dense Plasma

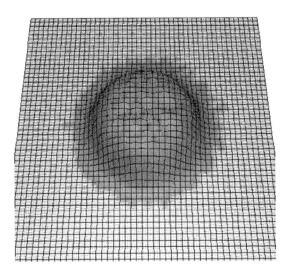

Diese Arbeit wurde als Dissertation an der
Humboldt-Universität zu Berlin zugelassen
und am 23. März 2004 erfolgreich verteidigt.

Bibliografische Information Der Deutschen Bibliothek

Die Deutsche Bibliothek verzeichnet diese Publikation in der Deutschen
Nationalbibliografie; detaillierte bibliografische Daten sind im Internet über
http://dnb.ddb.de abrufbar.

ISBN 3-8325-0569-5

Logos Verlag Berlin
Comeniushof, Gubener Str. 47,
10243 Berlin
Tel.: +49 030 42 85 10 90
Fax: +49 030 42 85 10 92
INTERNET: http://www.logos-verlag.de

Contents

List of Figures

1 Introduction

In plasmas of sufficiently high density or sufficiently low temperature the mean potential energy is of the same order or even larger than the mean kinetic energy. Such systems are called strongly coupled or non-ideal plasmas. The occurrence of strongly coupled plasmas in nature and experiment is manifold. Astrophysical objects such as brown dwarfs, the interior of white dwarfs, neutron star crusts, main sequence stars and the Jovian planets consist of strongly coupled plasmas, cf. [CHABRIER & SCHATZMAN 94]. In laboratory experiments strongly coupled plasmas can be generated by shock compression, strong laser pulses, z-pinch, ion beams, wire explosions [DA SILVA ET AL. 97, COLLINS ET AL. 98, KNUDSON ET AL. 01, THEOBALD ET AL. 96, BAUMUNG ET AL. 96, KLOSS ET AL. 96], in capillary discharges [KUNZE ET AL. 94] or in dusty plasmas [CHU & I 94]. Technical devices that involve dense plasmas are e.g. X-ray lasers [ROCCA ET AL. 94] and inertial confined fusion targets [LINDL ET AL. 92].

The physical properties of strongly coupled plasmas are determined by COULOMB interaction and quantum-mechanical uncertainty and exchange. As a consequence the border between the discrete and continuous part of the energy spectrum is lowered, spectral lines become broader and are shifted, the equation of state (EOS) and transport properties are modified, excitation processes are enhanced, pressure ionization and metallization are observed. Theoretical approaches to the description of strongly coupled plasmas are based upon the statistical physics of interacting many-particle systems. Due to the complexity of the problem only few general results like a virial theorem, energy bounds and stability properties are available, see [KRAEFT ET AL. 86, for a review]. Many analytical methods and simulation approaches concentrate on certain regions of coupling strength, degeneracy, degree of ionization, charge number, density and temperature or treat simplified model plasmas, see [EBELING ET AL. 76, KRAEFT ET AL. 86, EBELING ET AL. 91, ICHIMARU 94, REDMER 97, BONITZ & SEMKAT 03].

The foremost task of a theoretical description of a plasma is the characterization of the phases and the equation of state, as they are of fundamental importance to physicists and astrophysicists alike. Already hydrogen the most simple and most abundant element has a rich phase diagram cf. [MAO & HEMLEY 94, EBELING ET AL. 96a, for reviews] and despite considerable theoretical and experimental efforts it is still uncertain in several important areas. Recent experiments have probed the effect of strong shocks on liquid hydrogen and deuterium cf. [WEIR ET AL. 96, DA SILVA ET AL. 97, COLLINS ET AL. 98, TERNOVOI ET AL. 99, MOSTOVYCH ET AL. 00, KNUDSON ET AL. 01] and the theoretical determination correspondings HUGONIOTS has attracted much interest, see [LENOSKY ET AL. 97, ROGERS & YOUNG 97, BUNKER ET AL. 97a, ROSS 98, BEULE ET AL. 99a, MILITZER & CEPERLY 00, JURANEK ET AL. 01, BEULE ET AL. 01, JURANEK ET AL. 02, KNAUP ET AL. 03, MILITZER 03]. The particular troublesome region contains the dissociation/ionization transition of the dense semiconducting molecular/atomic liquid to the highly ionized plasma which shows metal-like electrical conductivity. The nature and properties of this transition are not simple to deduce, since it occurs in a region where both thermal and pressure effects are important. The main theoretical issue, whether it is a first-order *plasma phase transition* (PPT) or rather a continuous transition, has been the source of a considerable amount of research and controversy, cf. [NORMAN & STAROSTIN 70, EBELING & SÄNDIG 73, ROBNIK & KUNDT 83, DIENEMANN ET AL. 80, EBELING & RICHERT 85b, FÖRSTER ET AL. 92, SAUMON & CHABRIER 92, SCHLANGES ET AL. 95, REINHOLZ ET AL. 95, HOLMES ET AL. 95, ROSS 96, EBELING ET AL. 96a, KITAMURA & ICHIMARU 98, NELLIS ET AL. 98, MILITZER & CEPERLY 00, BEULE ET AL. 01, NELLIS 02, BONITZ ET AL. 03, EBELING & NORMMAN 03]. Furthermore it is not obvious if the dissociation will proceed directly to a fully ionized regime or instead include an intermediate atomic phase.

An equilibrium description may be insufficient for many plasmas, e.g. those generated by high energy deposition on short time scales. In many strongly coupled plasmas it is reasonable to proceed with evolution equations for the macroscopic variables like particle densities, temperature or degree of ionization [EBELING ET AL. 89, KREMP ET AL. 89, SCHLANGES & BORNATH 93], other situations are dominated by magnetic forces or radiation effects, see e.g. [MEYER ET AL. 00, RADTKE ET AL. 00]. In partially ionized dense plasmas the interplay between ionization/recombination processes and temperature evolution is of special interest as these processes result in the consumption/production of large amounts of kinetic energy [OHDE ET AL. 95, BEULE ET AL. 96, BORNATH ET AL. 98]. Much work has been devoted to the proper inclusion of interaction and many-particle effects into the macroscopic evolution equations [KREMP ET AL. 89, LEONHARDT & EBELING 93, BONITZ 90, SCHLANGES & BORNATH 93, OHDE ET AL. 97, BORNATH ET AL. 01] resulting in generalized rate and transport coefficients that introduce additional non-linearities into the evolution equations. It turns out that excitation rates are exponentially enhanced and that the area of strong coupling may involve collective effects like bistability, nucleation and front propagation [EBELING ET AL. 87, EBELING ET AL. 89, KREMP ET AL. 89, BONITZ 90, SCHLANGES & BORNATH 93, BEULE ET AL. 98b].

This work consists of two parts, the first part (chapter 2 to 5) deals with thermodynamics of partially ionized strongly coupled plasmas, while the second part (chapter 6) treats the kinetics of electron transitions in homogeneous and inhomogeneous dense plasma. The thermodynamic properties of strongly coupled plasmas and the possibilities of representing them by means of a PADÉ approximation within the chemical picture are introduced in the next chapter. This technique is then applied in chapter 3 to calculate the adiabatic equation of state in hydrogen which is important to astrophysical application. Furthermore the detailed composition of a carbon-hydrogen plasma mixture is determined over a wide range of densities and temperatures and possible implications for plasmas of capillary discharges are discussed. In chapter 4 the equation of state for hydrogen is calculated in the region of the plasma phase transition by constructing a model that combines new results from Monte-Carlo simulation of the dense fluid with the PADÉ approach for the highly ionized phase. Then the equation of state of shock compressed deuterium is obtained and compared with experimental data as well as other model calculations (chapter 5). Appendices A summarizes approximations formulas used in the PADÉ interpolation scheme.

In the second part (chapter 6) a stochastic method for treating the kinetics of electron ionization/recombination and excitation/deexcitation transitions is developed and used to study the propagation of ionization fronts and the related non-perturbative fluctuation effects. The stochastic description can be combined with temperature evolution equations to study the interplay between ionization/recombination processes and temperature evolution. Appendix B contains details on the algorithm used in the stochastic simulations, while appendix C introduces an efficient tool to evaluate the crucial qualities of the random number generators needed in these simulations.

2 Thermodynamic Functions

Since the first investigations for the electron gas [WIGNER 34, BOHM & PINES 53, GELL-MANN & BRUECKNER 56] many different methods have been developed for the calculation of thermodynamic potentials of many-body systems with COULOMB interaction. The thermodynamic properties of plasmas are well understood in several limiting cases and for simplified model systems like the one component plasma (OCP) [DEBYE & HÜCKEL 23, ABE 59, COHEN & MURPHY 69] and highly degenerate electron gas [WIGNER 34, GELL-MANN & BRUECKNER 56]. The SLATER-sum method [MORITA 59, EBELING ET AL. 76], GREENs functions [KRAEFT ET AL. 86], and FEYNMAN-KAC path-integral method [ALASTUEY & PEREZ 96] have proven to be especially useful for the calculation of density expansion. Other approaches are based on integral equation like hyper-netted chain (HNC) [ICHIMARU 94] and density-functional theory (DFT) [HOHENBERG & KOHN 64, KOHN & SHAM 65, MERMIN 65] that is especially useful for highly degenerate systems [PERROT & DHARMA-WARDANA 84, XU & HANSEN 98]. The THOMAS-FERMI theory can be considered as a special case of DFT and becomes exact as the ion charge goes to infinity, cf. [DREIZLER & GROSS 90].

Despite all these different approaches there is still a lack of reliable analytical results for large density and temperature regions. In order to overcome this situation various simulation methods have been developed. The most sophisticated include path-integral Monte Carlo (PIMC) [PIERLEONI ET AL. 94, MILITZER & CEPERLY 00, FILINOV ET AL. 03] and ab-initio molecular dynamics method (AIMD) [HOHL ET AL. 93, DHARMA-WARDANA & PERROT 02] based on density functional theory. The computational requirements of these methods confine the sample size and simulation time to fairly small values or limit the accessible density-temperature range. In order to overcome such limitations, more approximate methods such as wave-packet molecular dynamics [KLAKOW ET AL. 94a, KLAKOW ET AL. 94b, EBELING ET AL. 96b, EBELING & MILITZER 97, KNAUP ET AL. 03], effective potentials [EBELING & SCHAUTZ 97, ORTNER ET AL. 97] and tight binding potentials [LENOSKY ET AL. 97, LENOSKY ET AL. 00] have been used. For many applications even these more approximative simulation methods are still to time consumptive. Furthermore problems encountered in plasma physics often require the knowledge of the thermodynamic properties over a wide range of densities and temperatures while the range of validity of single simulation approaches or analytical results is limited. This problem may be overcome either by producing large tables [KERLEY 80, SAUMON ET AL. 95] or by the construction of interpolation formulas for the required properties. The interpolation schemes may be merely numerical [ICHIMARU 87, CHABRIER & POTEKHIN 98] or contain analytical representations of the limiting laws [EBELING & RICHERT 85b, EBELING 90, STOLZMANN & BLÖCKER 96, STOLZMANN & BLÖCKER 00]. In the area of partial ionization this approach can be combined with the chemical picture by constructing an expression for a thermodynamic potential, usually the free energy, in terms of the various chemical species. The ionization equilibrium is then determined by minimizing of the free energy function with respect to detailed composition [EBELING & RICHERT 85b, MARLEY & HUBBARD 88, SAUMON & CHABRIER 91, SAUMON & CHABRIER 92, POTEKHIN 96, BEULE ET AL. 99a]. This approach has been successfully applied in many different situations [EBELING ET AL. 91, STOLZMANN & BLÖCKER 96, BEULE ET AL. 01] and will be used in this work.

This chapter first gives a classifcation of plasmas by means of dimensionless parameter. Then the chemical picture is introduced in connection with the results of the low density expansions and the second virial coefficient. It follows a discussion of ionization equilibrium and partition sums before PADÉ formula for the COULOMB contributions are given. Finally the possibilities of representing the interaction of composite particles are discussed.

2.1 Plasma Parameter

Consider a plasma that consists of electrons with mass m_e and charge $q_e = -e$ and ions with mass m_i and charge $q_i = +Ze$ so that the plasma is charge neutral as a whole $ZN_e = N_i$. In order to characterize the plasma state dimensionless parameters, like degeneracy parameter θ and coupling parameter Γ can be used.

The coupling parameter measures the ratio between typical potential and kinetic energy and may be defined for electrons Γ_e and ions[1] Γ_i. For a classical plasma one defines:

$$\Gamma_e = \frac{e^2}{4\pi\varepsilon_0 d_e k_B T} \qquad \text{and} \qquad \Gamma_i = \frac{(Ze)^2}{4\pi\varepsilon_0 d_i k_B T} \qquad (2.1)$$

Here $d_i = (4\pi n_i/3)^{-1/3}$ denotes the ion-sphere radius. The corresponding quantity for the electrons $d_e = (4\pi n_e/3)^{-1/3}$ is usually called the WIGNER-SEITZ radius. Only for hydrogen-like plasmas ($Z = 1$) the electron and ion coupling parameter coincide, otherwise $\Gamma_i = Z^{5/3}\Gamma_e$. As long as the kinetic energy is significantly larger than the potential energy ($\Gamma \ll 1$) the interaction effects might be neglected (ideal plasma) or can be treated by perturbation theory (weakly coupled plasma). A strongly coupled plasma ($\Gamma \gtrsim 1$) refers to the case where the system begins to exhibit features that are qualitatively different from weakly coupled systems. A dimensionless parameter that measures the strength of the ion-electron interaction is given by

$$\xi_e = 2Z\sqrt{I/k_B T} \qquad (2.2)$$

where I denotes the ionization energy of hydrogen. For $\xi_e \geq 1$ the interactions are essential for microscopic scattering processes, i.e. the BORN approximation breaks down. Finally the BRUECKNER parameter $r_s = d_e/a_B$ defined as the ratio of the WIGNER-SEITZ radius and the BOHR radius a_B allows to distinguish low density $r_s < 1$ and high density region $r_s > 1$.

The degeneracy parameter θ is given by the quotient of the thermal energy $k_B T$ and the FERMI energy E_F

$$\theta = \frac{k_B T}{E_F} \qquad \text{with} \qquad E_F = \frac{\hbar^2 (3\pi^2 n_e)^{\frac{2}{3}}}{2m_e} . \qquad (2.3)$$

---

[1] The generalization for multi-component plasmas is given in 2.26.

If $\theta \gg 1$ the plasma is called non-degenerate or classical and BOLTZMANN statistics apply. If $\theta \lesssim 1$ the plasma is called degenerate and has to be treated using quantum mechanics and FERMI-DIRAC statistics. Due to their higher mass the ions will degenerate only at much lower temperatures than the electrons and thus they will be treated classically in this work. In the degenerate region the typical kinetic energy in (2.1) and (2.2) is no longer $k_B T$ but rather $2k_B T I_{3/2}(x)/(n_e \Lambda_e^3)$, where $x = \mu_e/(k_B T)$ denotes the dimensionless chemical potential and $\Lambda_e = h^2/\sqrt{2\pi m_e k_B T}$ is the thermal DE BROGLIE-wave length of the electrons. The FERMI integral $I_{3/2}(x)$ is given in appendix A.1. Therefore Γ_e and ζ_e become independent of temperature in the highly degenerate region $\theta \ll 1$.

Figure 2.1 shows typical natural and man-made nonrelativistic plasmas in the density-temperature plane together with lines of constant electron coupling parameter $\Gamma_e = 0.1, 1, 10$ and degeneracy parameter $\theta = 0.1, 10$. A further line indicates where $r_s = 1$.

2.2   Chemical Picture

Physical and chemical pictures provide alternative descriptions of plasmas. The only constituents in the physical picture are electrons and nuclei while all other particles, e.g., atoms, ions, and molecules are considered as composites build from electrons and nuclei. The great advantage of the physical picture is its structural simplicity: All matter, independent of its state of ionization is considered as a system of point charges with COULOMB interactions, cf. [KRAEFT ET AL. 86]. Masses, charges, and abundancies of nuclei are the only input of the theory. On the other hand the physical picture is commonly based on diagrammatic expansions which converge only at low densities. Thus one has to resort to additional assumptions in order to progress to higher densities.

The chemical picture [EBELING ET AL. 76] uses a much more complicated model of a plasma. In this model the atoms, ions, and molecules are treated as separate species. Therefore the constituents of the plasma are free electrons, free nuclei, ions, atoms, and molecules and all these species are treated on equal footing. The advantage of the chemical picture is that it is in many cases more appropriate for the description of real plasmas, cf. [EBELING ET AL. 91]. This reflects the simple fact, that in many respects atoms and molecules behave as point

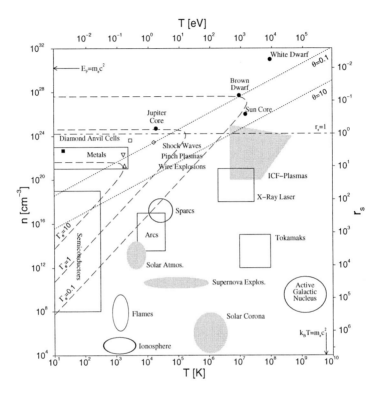

Figure 2.1: Overview of densities and temperatures ranges for selected natural and man made plasmas (adapted from [REDMER 97]). The plasma parameters are represented by lines of constant coupling parameter $\Gamma_e = 0.1, 1, 10$ (long dashed lines), degeneracy parameter $\theta = 0.1, 10$ (dotted lines) and BRUECKNER parameter $r_s = 1$ (dash-dotted line). The triangles mark the critical points of the metal-insulator transition of Hg ▽, and Cs △. The diamond ◇ indicates the predicted location of the PPT for hydrogen. Shock wave experiments in hydrogen usually start out form the molecular liquid (filled box ■), while the open square □ marks the experimental observation of the insulator-to-metal transition by [WEIR ET AL. 96].

masses/charges and the internal structure is rather irrelevant. The chemical picture combined with the free energy minimization method represents a reasonable compromise between rigorous treatment and practical applicability but it needs a careful quantum-statistical analysis in order to avoid double counting of effects. The free energy model is usually formulated in terms of the chemical species which dominate at low density.

Density Expansion: The SLATER-sum method [MORITA 59] has proven to be especially useful for the calculation of the second virial coefficent in weakly degenerate multi-component plasmas. This method maps quantum mechanical interaction onto classical potentials, that have a short range character. The resulting effective potentials allow the application of the classical cluster expansion. The method has been used to derive an analytic expression for the second virial coefficient for weakly degenerate ($\theta \gtrsim 1$) and weakly coupled ($\Gamma \gtrsim 1$) plasmas, cf. [EBELING 67a] to [EBELING 68a] and [EBELING ET AL. 76]. The method of GREENs functions was used by in order to obtain an analytic representation for the second virial coefficient, that is consistent with the results from the SLATER-sum method cf. [KRAEFT ET AL. 86]. Using the FEYNMAN-KAC path-integral method these results could be confirmed and an additional term (order $n^{5/2}$) could be determined [ALASTUEY & PEREZ 96, and references therein]. These expansions provide limiting laws for thermodynamic potentials at low densities. Furthermore they offer a systematic way to achieve the self consistent convergence of the internal partition function.

Free Energy Minimization: Once the HELMHOLTZ free energy functional F has been defined by means of expansions, PADÉ approximation or with the PACH approach (see section 2.8 the detailed composition of the plasma can be found. At given volume V and temperature T, the equilibrium state is determined by minimizing F with respect to the particle numbers of the chemical constituents $\{N_v\}$ subject to the stoichiometric constraints, see section 2.3.

$$F(T,V\{N_v\}) \rightarrow \text{Minimum} . \qquad (2.4)$$

Then pressure p, entropy S and related quantities are obtained from F using the well known thermodynamic relations. One major problem in calculating the particle numbers $\{N_v\}$ is to achieve the self consistent convergence of the internal partition function.

2.3   Ionization/Dissoziation Equilibrium

Consider a plasma contained in a volume V at fixed temperature T. The different constituents (index v) within the chemical pictures are characterized by their particle masses m_v, charge numbers z_v, and densities n_v. The mutual transformations between the different particles can always be summarized by a set of linear independent chemical reaction equations

$$\sum_v S_{v\tau} n_v = 0 \,, \tag{2.5}$$

where $S_{v\tau}$ denote the stoichiometric coefficients of the different reactions τ. Proper stoichiometric constraints $S_{v\tau}$ also guarantee that electro-neutrality and conservation of mass density ρ,

$$\sum_v z_v n_v = 0 \quad \text{and} \quad \sum_v m_v n_v = V\rho \,, \tag{2.6}$$

are always fullfilled. The discussion of ionization and dissociation equilibrium starts with a volume density of free energy

$$\frac{F}{V} = f(\{n_v\}, T) \,. \tag{2.7}$$

At constant volume and temperature it takes on a minimum in thermodynamic equilibrium. The corresponding variational principle $\delta f(\{n_v\}) = 0$ together with the constraints given by (2.5) leads to a set of coupled mass action laws or SAHA equations [SAHA 21]

$$\sum_v S_{v\tau} \frac{\partial f(\{n_v\})}{\partial n_v} = \sum_v S_{v\tau} \mu_v = 0 \,, \tag{2.8}$$

where $\mu_v = \partial f/\partial n_v$ denotes the chemical potential of species v.

Direct Optimization: For multi-component systems the set of coupled SAHA equations (2.8) can become difficult to solve due to their non-linear character generated by the mutual dependence of all μ_v and all particle densities via the interaction term $\mu_v^{int}(\{n_v\})$. Furthermore several solutions exist in regions of thermodynamic instability corresponding to several local minima of the free energy. Therefore it is often advantageous to treat the constrained optimization problems

$$\delta f(\{n_v\}) = 0 \quad \text{and} \quad \sum_v S_{v\tau} n_v = 0 \,, \tag{2.9}$$

directly instead of proceeding to (2.8). There exist several general and robust algorithms to solve constrained optimization problems, cf. [PRESS ET AL. 92]. Evolutionary algorithms [RECHENBERG 73, SCHWEFEL 81, EBELING & FEISTEL 82] and simulated annealing procedures [KIRKPATRICK ET AL. 83] are also capable of over coming the problem of local minima. For all problems encountered in this work simulated annealing with a simple step controlled multiplicative annealing scheme cf. [PRESS ET AL. 92] proved to be sufficient.

Non-Ideal Saha Equation: Lets concentrate on a single ionization equilibrium between z fold and $z + 1$ fold charged ion/atom described by the reaction equation and equilibrium condition

$$i_z \leftrightharpoons i_{z+1} + e \qquad \Longrightarrow \qquad \mu_z = \mu_{z+1} + \mu_e . \tag{2.10}$$

By separating the chemical potentials into ideal and interaction part $\mu_v = \mu_v^{id} + \mu_v^{int}$ one yields the non-ideal SAHA equation

$$\frac{n_{z+1}}{n_z} = \frac{\sigma_{z+1}}{\sigma_z} \exp\left(\frac{\mu_e^{id} + \Delta I_z}{k_B T}\right) . \tag{2.11}$$

The ions can be treated classically

$$\mu_z^{id} = k_B T \ln \frac{n_z \Lambda_z^3}{\sigma_z} , \tag{2.12}$$

where σ_z denotes the partition function of the z fold charged ions. The ideal chemical potential of the electrons μ_e^{id} is derived from FERMI-DIRAC statistics, cf. section 2.4.

In comparison to an ideal model plasma with no interaction the ionization energy of the z fold charged ion I_z is decreased by a shift

$$\Delta I_z = \mu_z^{int} - \mu_{z+1}^{int} - \mu_e^{int} , \tag{2.13}$$

that determined by the interaction contributions of the chemical potentials and leads to an increase in density n_{z+1} of the higher charged ions i_{z+1}.

Partition Sums: In equation one requires the partition sum of an atom/ion with charge z ionization energy I_z ($I_z > 0$) and bound states with binding energies E_z^m where $-I_z < E_z^m < 0$ that are g_z^m-fold degenerate. Within classical statistical physics the internal partition sum of this ion is

$$\sigma_z = \sum_m g_z^m \exp\left(\frac{E_z^m}{k_B T}\right) , \tag{2.14}$$

where g_z^m denotes the degeneracy of the bound states. It is well known, that (2.14) is divergent for hydrogen-like spectra [HERZFELD 16]. As the development of highly excited states $m \longrightarrow \infty$ and $E_z^m \to 0$ is disturbed by the presence of neighboring particles their weight has to be reduced by weight factors w_z^m

$$\sigma_z = \sum_m w_z^m g_z^m \exp\left(\frac{E_z^m}{k_B T}\right) , \qquad (2.15)$$

that results in cutting the highly excited states. Several scheme have been proposed [MARLEY & HUBBARD 88, LARKIN 60, EBELING & HILBERT 02]. In this work the convergence of the partition sum is guaranteed by using the PLANCK-BRILLOUIN-LARKIN (PBL) weight factors[2]

$$w_z^m = 1 - \exp\left(\frac{E_z^m}{k_B T}\right) + \frac{E_z^m}{k_B T}\exp\left(\frac{E_z^m}{k_B T}\right) , \qquad (2.16)$$

that are obtained naturally from quantum statistical consideration for weakly non-ideal hydrogen-like plasma, [EBELING ET AL. 91]. E_z^m denote the unshifted energy levels of the isolated ion/atom.

When considering multiple ionization stages with charge numbers $z = 0,\ldots,Z$ simultaneously the situation is slightly more complex as the partition sums have to be determined consistently with respect to a single reference energy. Choosing the energy of the naked core (or the most highly charged ion) to be zero results in:

$$\sigma_Z = g_Z , \qquad (2.17)$$

$$\sigma_z = \sum_m g_z^m \left[\exp\left(\frac{-E_z^m}{k_B T}\right) - 1 + \frac{E_z^m}{k_B T}\right]\exp\left(\frac{\sum_{k=z+1}^{Z-1} I_k}{k_B T}\right) , \quad z = 0,\ldots,Z-1 .$$

Molecules: In low temperature plasmas atoms/ions can form molecules. Rotational and vibrational excitations of molecules are additional internal degrees of freedom that have to be included into the partition sum. E.g. for diatomic H_2 molecules one may use the following expression

$$\sigma_{H_2} = \frac{16\exp\left(\frac{2I}{k_B T}\right)\exp\left(\frac{D_0}{k_B T}\right)\exp\left(\frac{-hcB}{2k_B T}\right)\frac{k_B T}{hcB}}{1 - \exp\left(\frac{-hc\omega}{k_B T}\right)} . \qquad (2.18)$$

---

[2]When using the occupation numbers of excited states obtained within the PBL approach for calculation of optical properties a rescaling is required, cf. [ROGERS 86].

with rotational $B_{H_2} = 60.853\,\mathrm{cm}^{-1}$ and vibrational $\omega_{H_2} = 4401\,\mathrm{cm}^{-1}$ energy constants and binding energy $D_{0,H_2} = 4.735\,\mathrm{eV}$, cf. [HUBER & HERZBERG 79]. For partition sum of deuterium molecules rotational and vibrational constants are $B_{D_2} = B_{H_2}/2$ and $\omega_{D_2} = \omega_{H_2}/\sqrt{2}$ due to mass scaling. The approximation (2.18) requires $T \gg T_r = hcB/k_B \approx 85\,\mathrm{K}$ and excited electronic states as well as the weak non-linearities in the molecular force constants that also result in coupling between rotational and vibrational excitation are neglected.

2.4　Free Energy Model

The central assumption of the free energy minimization approach is the factorization of the many-body partition function in translational, internal and interaction factors corresponding to the separation of the HELMHOLTZ free energy F.

The free energy density expression used here for a two-component system of neutral f_0 and charged particles $f_\pm$,

$$f(V,T,n_\mathrm{v}) = f_0(V,T,n_\mathrm{v}) + f_\pm(V,T,n_\mathrm{v}), \qquad (2.19)$$

combines results for the fully ionized plasma domain [EBELING & RICHERT 85a, EBELING 90, STOLZMANN & BLÖCKER 96, STOLZMANN & BLÖCKER 00] with data for the dense, neutral fluid calculated within a dissociation model [FÖRSTER 92, BUNKER ET AL. 97a]. Both contributions to the free energy are split into ideal and interaction parts

$$f = f^{id} + f^{int}. \qquad (2.20)$$

Interactions in the neutral and in the charged subsystem, are taken into account seperately, whereas the interaction between charges and neutrals is accounted for by the reduced-volume concept, see section 2.8. Similar approaches have been used by [ROGERS 86, SAUMON & CHABRIER 91, SAUMON & CHABRIER 92, POTEKHIN 96, CHABRIER & POTEKHIN 98, JURANEK ET AL. 01, EBELING ET AL. 03].

Ideal Contributions: The free energy density of an ideal electron gas is exactly know and can be represented by means of FERMI integrals (see appendix A.1)

$$f_e^{id} = \frac{2k_BT}{\Gamma_e}\left[x_e I_{\frac{1}{2}}(x_e) - I_{\frac{3}{2}}(x_e)\right] \quad \text{where} \quad x_e = \frac{\mu_e^{id}}{k_BT}, \qquad (2.21)$$

denotes the dimensionless ideal chemical potential of the electrons determined by equation A.3

$$I_{\frac{1}{2}}\left(\frac{\mu_e^{id}}{k_B T}\right) = \frac{n_e \Lambda_e^3}{2} \,. \tag{2.22}$$

In order to calculate the free energy density for a given density and temperature one needs to invert equation (2.22) first. However, in practical calculations one prefers analytical expressions in terms of density and temperature. For this purpose several parametrizations of the inverse of $I_{1/2}$ have been constructed [PERROT & DHARMA-WARDANA 84, ZIMMERMANN 88]. The later one is given in appendix A.1 and is used throughout this work together with the corresponding parametrization of the FERMI integrals.

For the contribution of the ions it is reasonable to neglect the degeneracy effects. For a mixture of differently z-fold charged ions one yields

$$f_i^{id} = k_B T \sum_z n_z \left(\ln \frac{n_z \Lambda_z^3}{\sigma_z} - 1\right), \tag{2.23}$$

where n_z denotes the density of the z-fold charged ions and the partition sums σ_z have been given in 2.17 An analogous formula applies for the contribution of atoms and molecules using the partition sums introduced in the previous section.

2.5   Charged Particle Subsystem

For the COULOMB part of the free energy density efficient approximation formulas have been suggested by [EBELING & RICHERT 85a, EBELING 90] using the PADÉ interpolation method. The method is applicable to multi-component plasmas including multiple charged ions and uses the following decomposition:

$$f_\pm^{int} = f_e + f_i + f_{ie} \,. \tag{2.24}$$

The contributions f_e and f_i account for the correlations and the exchange in the electronic and ionic subsystem, while f_{ie} describes the electron-ion screening.

The PADÉ approach is based on analytical results for the quantum virial expansion [EBELING ET AL. 76] and discussed in detail in [EBELING ET AL. 91]. At low densities and/or high temperatures the quantum corrected DEBYE law is reproduced. With increasing density, ions and electrons behave in quite different ways. The ions form a subsystem of classical, strongly coupled particles with a lattice-like structure screened by a partially degenerate electron liquid. There, the PADÉ approximation converges to the results of respective Monte Carlo simulations for the ion-ion and ion-electron interactions. An improved expression for the electronic contribution f_e was given by [STOLZMANN & BLÖCKER 96, STOLZMANN 96] through considering additional thermal corrections.

Ion Contributions: For the ion-ion contribution one can take advantage of the PADÉ formula proposed by [EBELING 90, EBELING ET AL. 91]. For the densities considered here the ionic quantum correction supposed by [STOLZMANN & BLÖCKER 96] can be neglected.

$$f_i = -n_+ k_B T \frac{b_0 \Gamma_i^{\frac{3}{2}} + b_2 \Gamma_i^{\frac{9}{2}} \varepsilon_i}{1 + b_1 \Gamma_i^{\frac{3}{2}} + b_2 \Gamma_i^{\frac{9}{2}}}, \tag{2.25}$$

where

$$n_+ = \sum_i n_z, \qquad \Gamma_i = \frac{e^2}{4\pi k_B T} \left(\frac{4\pi n_+}{3} \right)^{\frac{1}{3}}, \qquad \tau = \frac{k_B T}{\text{Ry}} \tag{2.26}$$

denotes the density of ions, the ionic coupling constant for ionic mixtures and the temperature in atomic units, respectively. The weak coupling limit ($\Gamma_i \ll 1$) is determined by the Debye OCP law and ε_i denotes the classic ionic contribution in the strong coupling limit as calculated by [STRINGFELLOW ET AL. 90], see appendix A.2. The coefficients are given by

$$b_0 = \frac{1}{\sqrt{3}} \frac{\langle z^2 \rangle^{\frac{3}{2}}}{\sqrt{\langle z \rangle} \langle z^{\frac{3}{}} \rangle^{\frac{3}{2}}}, \qquad b_2 = 1000 \left(\frac{3}{4\pi} \right)^{\frac{3}{2}},$$

$$b_1 = \frac{3}{4\pi} \frac{\sqrt{\tau}}{b_0 \langle z \rangle \langle z^{\frac{5}{3}} \rangle^3} b_4, \qquad \langle z^a \rangle = \sum_i \frac{n_i z_i^a}{n_i}, \tag{2.27}$$

$$b_4 = \frac{\pi^{\frac{3}{2}}}{8} \sum_{i,j} \zeta_i \zeta_j z_i^2 z_j^2 \sqrt{\gamma_i + \gamma_j} - 2\pi\tau \sum_i \zeta_i^2 \frac{-1^{2s_i}}{2s_i + 1} \left(\frac{\gamma_i}{2} \right)^{\frac{3}{2}} E_2 \left(-z_i^2 \sqrt{\frac{2}{\gamma_i \tau}} \right)$$

$$\gamma_i = \frac{m_e}{m_i}, \qquad \zeta_i = \frac{n_i}{n_+}, .$$

For hydrogen we have only one ion species with charge one and therefore the above coefficients simplify significantly

$$b_0 = \frac{1}{\sqrt{3}}, \qquad b_1 = \frac{3}{4\pi} \frac{\tau}{b_0} \left[\frac{\pi\sqrt{2}}{8} - \frac{\pi\tau}{2\sqrt{2}} E_2 \left(-\sqrt{2/\tau} \right) \right]. \tag{2.28}$$

An interpolation formula for the quantum-virial function E_2 is introduced in section 2.6.

Electron Contributions: For the electron gas exchange and correlation effects have to be considered. In the limit of zero temperature the exchange and correlation contribution to the thermodynamic functions are comparatively well understood. Low [WIGNER 34, BAUS & HANSEN 80] and high density limits [GELL-MANN & BRUECKNER 56] are know exactly and diffusion Monte-Carlo simulations [CEPERLEY & ALDER 80] can be utilized to fill the gap.

$$f_e = -n_e k_B T \frac{f_0 \Gamma_e^{\frac{3}{2}} + f_6 \Gamma_e^3 - f_2 \Gamma_e^6 \tau^{-1} \varepsilon}{1 + f_1 \Gamma_e^{\frac{3}{2}} + f_2 \Gamma_e^6} \qquad (2.29)$$

$$f_0 = \sqrt{\frac{1}{3}} f_2 = 3, \qquad f_3 = \frac{1}{4}\sqrt{\frac{\tau}{\pi}},$$

$$f_1 = \frac{\sqrt{2}}{8 f_0}\left[1 + \frac{2\tau}{\sqrt{\pi}} E_2\left(-\sqrt{\frac{2}{\tau}}\right)\right], \qquad \tau = \frac{k_B T}{\text{Ry}},$$

where r_s denotes the BRUECKNER parameter and $\tau = k_B T / \text{Ry}$ the temperature in reduced units. ε denotes the ground state energy of the electron gas as given by equation (A.10).

Alternatively the electron-electron interaction contributions to the free energy can be split into HARTREE-FOCK exchange f_x and correlation energy f_c

$$f_{xc} = f_x + f_c. \qquad (2.30)$$

Exchange: The HARTREE-FOCK exchange contribution is exactly known for all densities and temperatures. It is given by

$$f_x = -\frac{e^2}{2\pi\varepsilon_0 \Lambda_e^4} \int_{-\infty}^{y} I_{-1/2}^2 \, \mathrm{d}y \qquad \text{with} \qquad y = \frac{\mu_e^{id}}{k_B T}, \qquad (2.31)$$

where $I_{-1/2}$ denotes a FERMI integral cf. section A.1. Limiting case $T = 0$ and low density or/and high temperatur can be evaluated explicitly. Otherwise numerical evalution methods or interpolation schemes have to be used, see [STOLZMANN 96] for discussion.

Correlation: By seperating the exchange and correlation parts [STOLZMANN & BLÖCKER 96] were able to introduce finite temperature correction in the correlation part

$$f_e^c = -n_e k_B T \frac{a_0 \Gamma_e^{\frac{3}{2}} - a_2 \Gamma_e^6 \left[\varepsilon_c(r_s) + \Delta\varepsilon_c(r_s,\tau)\right]/\tau}{1 + a_1 a_3 \Gamma_e^{\frac{3}{2}} + a_2 \Gamma_e^6} \tag{2.32}$$

$$a_0 = \frac{1}{\sqrt{3}} f_0(\Gamma_e), \qquad f_0(\Gamma_e) = \frac{1}{2}\left(\frac{1}{\sqrt{1+0.1088\Gamma_e}} + \frac{1}{(1+0.3566\Gamma_e)^{\frac{3}{2}}}\right)$$

$$a_1 = \frac{3\sqrt{3}}{32}\sqrt{2\pi}\,\tau\left[1 + \frac{2\tau}{\sqrt{\pi}} E_2\left(-\sqrt{\frac{2}{\tau}}\right)\right], \qquad a_2 = 6\frac{\tau}{r_s^4}, \qquad a_2 = \frac{1}{1+1.5\Gamma_e^{\frac{3}{4}}},$$

where $\Delta\varepsilon$ is given by

$$\Delta\varepsilon = 8\left(\frac{4}{9\pi}\right)^{\frac{4}{3}}\frac{r_s^2}{\Gamma_e^2}\frac{r_s + 1.05 r_s^{-0.35}}{1 + 1.5 r_s + r_s^2}\ln\left[2\left(\frac{4}{9\pi}\right)^{\frac{2}{3}}\frac{r_s}{\Gamma_e}\right]. \tag{2.33}$$

While (2.32) is expected to be more precise in the intermediate region (2.29) proves to be more suitable when higher numerical derivatives are required in order to obtain further thermodynamic properties.

Ion-Electron Contributions: Finally one has to include the screening effects between electrons and ions. The PADÉ approximation was proposed in [EBELING 90]

$$f_{ie} = -\frac{c_0 \Gamma_i^{\frac{3}{2}} + c_2 \Gamma_e^{\frac{9}{2}} e_{ie}}{1 + c_1 \Gamma_i^{\frac{3}{2}} + c_2 \Gamma_e^{\frac{9}{2}} + 2c_4 \Gamma_i^{\frac{3}{2}}\ln\left(1 + \sqrt{c_5/\Gamma_i^9}\right)}. \tag{2.34}$$

The coefficients are given by

$$c_0 = \frac{1}{\sqrt{3}}\frac{(\zeta + \langle z^2\rangle)^{\frac{3}{2}} - \zeta^{\frac{3}{2}} - \langle z^2\rangle^{\frac{3}{2}}}{\langle z\rangle^{\frac{1}{2}} + \langle z^{\frac{5}{3}}\rangle^{\frac{3}{2}}} \qquad c_2 = \left(\frac{3}{4\pi}\right)^{\frac{3}{2}}\pi^{\frac{9}{4}}\tau^{\frac{9}{4}} \tag{2.35}$$

$$c_1 = \frac{3}{4\pi}\frac{\sqrt{\tau}}{c_0\langle z\rangle\langle z^{\frac{5}{3}}\rangle^3}\cdot$$

$$\left\{\frac{\pi^{\frac{3}{2}}}{4}\zeta\langle z^2\rangle - \frac{2\pi}{\sqrt{\tau}}\sum_a\sum_b \zeta_a\zeta_b z_a^3 z_b^3\left(\frac{C_3(-|\xi_{ab}|)}{|\xi_{ab}|^3} - \frac{1}{6}\ln(|\xi_{ab}|) - \frac{C_E}{3} + \frac{11}{36}\right)\right\}$$

$$c_4 = \frac{1}{8c_0}\frac{(\langle z^3\rangle - \zeta)^2}{\langle z\rangle\langle z^{\frac{5}{3}}\rangle^3}, \qquad c_5 = \frac{4\pi}{3}\langle z\rangle\langle z^{\frac{5}{3}}\rangle^3\exp c_6$$

$$c_6 = \frac{\langle z^3 \rangle - \zeta}{z^3 \rangle - \zeta} - \ln\left(29.09(\langle z^2 \rangle + \zeta)\right) , \quad \xi_{ab} = -\frac{2z_a z_b}{\sqrt{\tau(\gamma_a + \gamma_b)}} , \quad \zeta = \frac{n_e}{n_+}$$

where the function $C_3(x)$ is discussed in section 2.6 and the strong coupling limit is given by

$$\varepsilon_{ie} = \frac{1}{1 + 1.137 r_s \sqrt{\tau}} \frac{0.76633 \, \Gamma_i^{1/4} r_s}{1 + 0.35325 \, \Gamma_i^{-1/4}} + \frac{\Gamma_i r_s (0.04504 + 0.00999 \, r_s)}{1 + 0.0887 \, r_s^2} . \quad (2.36)$$

This PADÉ formula interpolates screening corrections to the OCP from [GALAN & HANSEN 76] with a high density expansion from [KAGAN ET AL. 77]. The weak coupling limit Γ_i is given by DEBYE-HÜCKEL approximation of the two component plasma.

2.6   Quantum Virial Functions

The quantum virial functions used in the previous section are defined by the convergent series [EBELING ET AL. 76]

$$Q_3(-x) = \frac{x^3}{6}\left(\frac{C_E}{2} + \ln 3 - \frac{1}{2}\right) + \sum_{n=4}^{\infty} \frac{\sqrt{\pi}\,\zeta(n-2)}{\Gamma\left(\frac{n}{2}+1\right)}\left(-\frac{x}{2}\right)^n \quad (2.37)$$

and

$$E_2(-x) = \frac{x^2 \sqrt{\pi}\ln 2}{4} + \sum_{n=3}^{\infty} \frac{\sqrt{\pi}\,\zeta(n-1)(1-2^{2-n})}{\Gamma\left(\frac{n}{2}+1\right)}\left(-\frac{x}{2}\right)^n \quad (2.38)$$

where $C_E \approx 0.577216$ denotes the EULER constant, $\Gamma(n)$ the Gamma function, $\zeta(n)$ the RIEMANN Zeta function and the argument can be rewritten as

$$x = -\frac{2z_a z_b}{\sqrt{\tau(\gamma_a + \gamma_b)}} , \quad \tau = \frac{k_B T}{\text{Ry}} , \quad \gamma_a = \frac{m_e}{m_a} . \quad (2.39)$$

Note that $E_2(x)$ considers only equal particle $z_a = z_b$ leading to $x = -z_a^2 \sqrt{2/(\tau\gamma_a)}$ for ions and $x = -\sqrt{2/\tau}$ for electrons. After introducing

$$Q_3^*(-x) = \frac{Q_3(-x)}{x^3} , \quad \text{and} \quad E_2^*(-x) = \frac{4}{\sqrt{\pi}\,x^2} E_2(-x) \quad (2.40)$$

the following expansions for $x \ll 1$, i.e. high temperatures are valid

$$Q_3^*(-x) = \frac{1}{6}\left(\frac{C_E}{2} + \ln 3 - \frac{1}{2}\right) + \frac{\pi^{\frac{5}{2}}x}{192} - \frac{x^2}{50} , \quad (2.41)$$

$$E_2^*(-x) = \ln 2 - \frac{\pi^{\frac{3}{2}}x}{18} + 0.113 x^2 . \quad (2.42)$$

At low temperatures, i.e. $x \gg 1$ the evaluation of the sum representation is disadvantageous but the following expansions can be used [EBELING 68d, STOLZMANN & BLÖCKER 96]

$$Q_3^*(-x) = \frac{1}{6}\left(\ln|3x| + 2C_E - \frac{11}{6}\right) + \frac{\sqrt{\pi}}{8x} - \frac{1}{12x^2}, \qquad (2.43)$$

$$E_2^*(-x) = \frac{2}{\sqrt{\pi x}} - \frac{1}{x^2}. \qquad (2.44)$$

For efficient numerical treatment the infinite sums are approximated by PADÉ formulas [BEULE ET AL. 98a] that are improvements of those given in [STOLZMANN 96]. When interpolating $Q_3^*(-x)$ with a PADÉ approximation

$$Q_3^*(-x) = \frac{0.14787 + \frac{\pi^{\frac{5}{2}}x}{192} + \frac{1.12x^{5.1}}{6}\left(\ln|3x| + 2C_E - \frac{11}{6} + \frac{3\sqrt{\pi}}{4x} - \frac{1}{2x^2}\right)}{1. + 1.12x^{5.1}}, \qquad (2.45)$$

it is useful to omit the $-x^2/50$ term of the high temperature expansion (2.41) for the sake of a smooth interpolation. In Figure 2.2 this interpolation formula is compared with the limiting laws (2.41) and (2.43) and the sum representation (2.37), where the first 100 terms were used. For $E_2^*(-x)$ the PADÉ approximation

$$E_2^*(-x) = \frac{\ln 2 - \frac{\pi^{\frac{3}{2}}x}{18} + 0.133x^2 + 0.14x^{5.9}\left(\frac{2}{\sqrt{\pi x}} - \frac{1}{x^2}\right)}{1 + 0.14x^{5.9}} \qquad (2.46)$$

includes all terms of the limiting laws (2.42) and (2.44).

A comparison of (2.46) with the first 100 terms of the sum representation (2.38) as well as the limiting laws (2.42) and (2.44) is shown in Figure 2.3. The excellent agreement between the PADÉ formula and the sum representation is reached by fitting the function that does the switching between the limiting laws i.e. $1.12x^{5.1}$ and $0.14x^{5.9}$, respectively.

2.7   Neutral Particle Fluid

So far mainly systems of point-like particles were considered. However, at small distances between composite particles the bound-electron shells become essential and lead to a strong repulsion between molecules, atoms and ions due to the

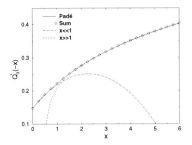

Figure 2.2: Quantum virial function $Q_3^*(-x)$ as obtained from the sum (2.37) (diamonds) the limiting laws (2.41) and (2.43) (dashed lines) and the PADÉ approximation (2.45) (solid line).

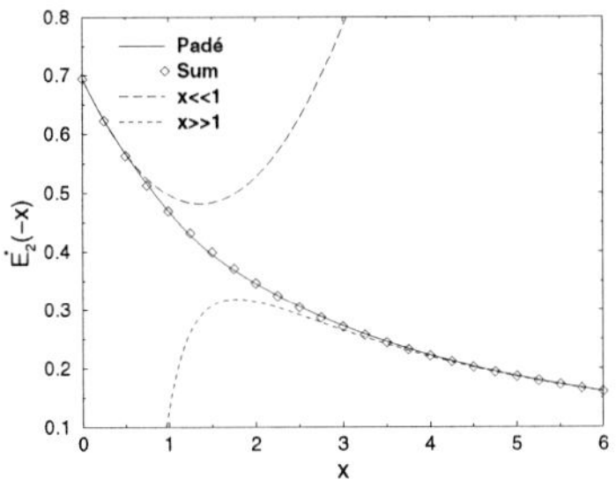

Figure 2.3: Quantum virial function $E_2^*(-x)$ as obtained from the sum (2.38) (diamonds) the limiting laws (2.42) and (2.44) (dashed lines) and the PADÉ approximation (2.46) (solid line).

PAULI exclusion principle. There are substantial areas of the density-temperatures plane where atoms or molecules dominate the detailed compositon and an appropriate description is provided by methods from fluid theory. The equation of state for neutral fluids can be determined applying various methods of liquid state theory such as integral equation techniques, classical Monte Carlo simulations, and fluid variational theory cf. [BUNKER ET AL. 97a, BUNKER ET AL. 97b]. In the fluid variational theory the equation of state is calculated by minimizing a free energy functional

$$f_0^{int} = f_{hc} + nk_Bt S(\eta) + \frac{n^2}{2} \int\limits_r^\infty \phi(r) g_{hs}(r,\eta) \, dr \tag{2.47}$$

with respect to the filling factor $\eta = \pi n d^3 / 6$. The reference system of hard spheres of diameter d and density n is modified by a soft-sphere correction which leads to good agreement with Monte-Carlo results.

The free energy of the hard sphere system f_{hc} given in appendix A.4 can be used an a first approximation of the interaction in neutral particle fluid. For partially dissociated dense hydrogen fluid, i.e. a mixture of H_2 molecules and H atoms, Monte Carlo simulations based on two-body potentials can be used to obtain an improved treatment, see section 4.1.

2.8 PACH Approach

The PACH (PADÉ Approximation in the Chemical Picture) approach combines
the methods discussed in the previous sections to obtain a free energy model for
partially ionized plasmas in terms of different chemical species. This model is
then used in conjunction with the free energy minimization technique introduced
in section 2.3 to obtain the detailed composition.

At high densities the COULOMB interaction of the bound outer shell electrons
of atoms or ions with surrounding free charged particle (electrons or ions) or po-
larisable free particles (atoms or molecules) leads to a decrease in binding energy.
For densities comparable to solid state densities the decrease is enhanced through
quantum mechanic exchange between bound electrons of neighboring particles.
Together with the decrease in binding energy caused by COULOMB interaction effects this leads to a sharp rise of the degree of ionization in plasma of high densities
i.e. at high pressure and is therefore called pressure ionization. The mechanism of
pressure ionization differs from usual ionization due to temperature. The average degree of ionization increases and is higher than predicted by a conventional
model that neglects such interactions. In the asymptotic case of very high densities
the degree of ionization in real high pressure plasma is completely different from
that in ideal model plasma (where interactions are neglected). The high pressure
plasma is completely ionized and consists only of free cores and free electrons,
while the degree of ionization in ideal model plasma tends towards zero and the
charged particles are all bound to neutral atoms. Here the reduced volume concept
is used to account for the interaction between the neutral and charged subsystem,
while polarisation effects are neglected.

Reduced Volume: It is interesting property of the equation of state of hard
spheres, that it also allows to consider point-like particles. The interaction between
point-like particles and the hard spheres can be rewritten as a reduction of the
volume that is accessible for the point-like particles [EBELING & RICHERT 85c]

$$V_{eff} = V(1 - \eta) \, , \tag{2.48}$$

where η denotes the packing parameter (A.13) and V_{eff} the effective accessible volume for the electrons. The reduced volume concept can be extended to degenerate point-like particles, see [FÖRSTER 92]. The resulting ideal free energy contributions for the electrons is

$$f_e^{id^*} = -n_e k_B T \frac{g_e}{n_e^* \Lambda^3} I_{\frac{3}{2}}\left(\frac{\mu_e^*}{k_B T}\right) - n_e \mu^* , \qquad I_{\frac{1}{2}}\left(\frac{\mu_e^{id^*}}{k_B T}\right) = \frac{n_e \Lambda_e^3}{(1-\eta)g_e} , \qquad (2.49)$$

and the interaction contributions f_e are rescaled accordingly, i.e. in formular (2.29) and (2.32) n_e is replaced by $n_e^* = n_e/(1-\eta)$. As a concequence free energy, chemical potential, and pressure of electrons are no longer a function of electron density only but also a function of the densities of all composite particles, i.e. the degree of ionization.

3 Temperature Ionization

Two applications of the methods introduced in the previous chapter are discussed next. In the first application the PACH approach is used to determine the detailed plasma composition during a capillary discharges. Spectroscopic data indicate that fully ionized carbon ions occur in capillary discharge experiment using a polyethylene tube. On the basis of the simultaneously obtained data for electron density and temperature we calculate the charge-state distribution for this multi-compenent plasma. For typical densities of the discharge plasma a temperature of $\geq 30\text{eV}/k_B$ is shown to be sufficient to generate a substantial amount of fully ionized carbon.

As a second application the adiabatic equation of state for partially ionized hydrogen plasma is determined. The knowledge of the adiabatic equation of state is required to describe the expansion of plasma or gas into vacuum, sound waves, ballistic-compressors, weak shocks and the structure of jovian planets. Bound-state formation significantly alters the adiabatic EOS and therefore an appropriate description of the adiabatic equation of state requires a proper determination of the ionization equilibrium.

3.1 Highly Ionized Carbon in Capillary Discharge Plasma

Since the sixties capillary discharge experiments [BOGEN ET AL. 65, CONRADS 67, BOGEN ET AL. 68] were of interest as highly reproducible light sources. Sparks emitting intense continuum radiation in the visible and ultraviolet spectral region are important for many applications: temperature determination in plasma by means of line-reversal method, for measurement of absorption coefficients in the ultraviolet and soft X-ray domains, especially of the discontinuities at the absorption edges, or for use as radiation standards. Capillary discharges are also considered as promising X-ray laser sources [ROCCA ET AL. 94, SHIN ET AL. 94]. Spectroscopic data indicate that fully ionized carbon ions occur in capillary discharge experiment using a polyethylene tube [CONRADS 67, SHIN ET AL. 94]. It is yet not fully understood which processes generate these ions although it was suggested that charge-exchange impacts play an important role [KUNZE ET AL. 94].

Experimental Situation: It is long known [GLEICHAUF 51] that discharges between two electrodes bridged by an insulator in high vacuum start in flashing over the insulating surface. A so-called sliding spark develops. Bridging the electrodes by a capillary, the same phenomenon was observed [BOGEN ET AL. 65, CONRADS 67, SHIN ET AL. 94, KUNZE ET AL. 94]. The capillaries were typically made of polyethylene in order to observe line-free recombination spectra of C^{5+}. In the initial phase of a capillary-discharge process the absorption of a large amount of energy by a relatively small amount of wall material generates a hot and dense plasma. In high-vacuum discharges the capillary is filled in much less than 50 ns. During the discharge the capillary wall material is evaporated and ionized due to JOULE heating and a hot dense carbon-hydrogen plasma is formed. Typical plasma parameters of capillary discharges are 10^{18} to 10^{21} cm^{-3} for electron density and 10^4 to 10^6 K for electron temperature.

Electron density and temperature in the plasma column can be determined by measurement of emitted absolute radiation intensity from optical thin and optical thick layers. Figure 3.1 shows these properties for a polyethylene capillary of 10 cm length and inner diameter of 2 mm [CONRADS 97].

As the process of ablation continues more and more material is released from the wall of the capillary and transformed into plasma. The increase in plasma density is accompanied by a steady decrease in temperature in the way that the pressure remains almost constant. Radiation from prominent bound-bound transitions allows to record the temporal appearance of ions of different charges. At the temperatures under consideration hydrogen is fully ionized. The experimental data of Figure 3.2 suggest that the plasma composition in the first few hundred nanoseconds is dominated by C^{5+} and C^{4+}. The carbon ions are identified by the line radiation from the transitions between the first excited states and the ground

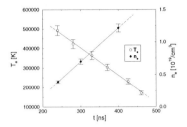

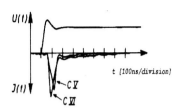

Figure 3.1: Experimental data for electron temperature T_e and electron density n_e as function of time by [CONRADS 97]. A polyethylene capillary of 10 cm length and inner diameter of 2 mm was used.

Figure 3.2: Temporal development of intensity $I(t)$ emitted by the C^{5+} line at 3.373 nm and the C^{4+} line at 4.027 nm and of discharge voltage $U(t)$, by courtesy of [CONRADS 97].

states, respectively.

Theoretical Modeling: The theoretical description aims at calculating the plasma composition over time. In the considered state of the discharge process recombination takes place by fast capture of electrons into highly excited levels which are followed by radiative de-excitation processes. The treatment [FÖRSTER ET AL. 98] is based on the assumption that the evolution of the charge-state distribution in the plasma column during the discharge process may be approximately described as a sequence of equilibrium states. The difference between ion and electron temperature is neglected. The detailed modeling of the excitation and de-excitation processes would need a more sophisticated treatment based on rate equations, cf. [SAWADA & FUJIMOTO 95, BEULE ET AL. 95, BORNATH ET AL. 98].

The equilibrium composition is determined using the PACH approach, i.e. by minimizing the free energy density

$$f(\{n_z\}, n_e, T) = f_i^{id} + f_e^{id} + f_i + f_e + f_{ie} + f_{hs}, \qquad (3.1)$$

with respect to the densities of the different ions $\{n_z\}$ and the free electron density n_e, see section 2.8. For these multi-component systems the correlation and exchange contributions f_e, f_i, and f_{ie} are calculated from (2.29), (2.25), and

(2.34) respectively. The ideal parts f_e^{id} and f_i^{id} are given by equations (2.21) and (2.23). The internal partition functions (2.14) of the ions are constructed according to the BRILLOUIN-PLANCK-LARKIN procedure (2.17) using spectroscopic data for the excited energy levels [MOORE 93]. Finally the hard-sphere radii r_z of the carbon ions with charge z, which are required for the short-range repulsion contribution f_{hs} equation (A.12) are chosen guided by the SLATER rule, cf. [BESPALOV & POLISHCHUK 89, EBELING ET AL. 91, section 4.4]

$$r_z = \frac{a_B}{0.172\sqrt{I_z\,[\text{eV}]}}\,, \qquad z = 0,\dots,Z-1\,, \qquad (3.2)$$

where I_z denotes ionization energy in eV and a_B denotes the BOHR radius.

Table 3.1: Radii of the different carbon ions in units of 10^{-10} m.

ion	C^{6+}	C^{5+}	C^{4+}	C^{3+}	C^{2+}	C^{1+}	atom
radius [10^{-10} m]	≈ 0	0.14	0.16	0.38	0.44	0.62	0.92

Detailed Composition: Figure 3.3 [FÖRSTER ET AL. 98] presents four maps which characterize the occurrence of C^{6+}, C^{5+}, C^{4+} and C^{3+} ions as part of a $(CH_2)_n$ plasma in the temperature range of 10^5 to 10^6 K and a mass-density range of 10^{-6} to $1\,\text{g cm}^{-3}$. These maps are obtained by solving the minimization problem and thus getting the detailed plasma composition on a dense grid of heavy-particle density n and temperature T.

The knowledge of the detailed composition implies naturally also the values of the free-electron density in any point of the ionization map. The experimental points of Figure 3.1 for the electron density and temperature may be in good approximation [FÖRSTER ET AL. 98] interpolated by linear functions of time t:

$$
\begin{aligned}
n_e(t) &= t \cdot 5.2\,10^{16}\,\text{cm}^{-3}\,\text{ns}^{-1} - 8.6\,10^{18}\,\text{cm}^{-3}\,, \qquad (3.3)\\
T_e(t) &= -t \cdot 1400\,\text{K ns}^{-1} + 823000\,\text{K}\,.
\end{aligned}
$$

The corresponding path for the evolution of the wall material during the discharge process is indicated by the full line in Figure 3.4 [FÖRSTER ET AL. 98]. Time t plays now the role of a parameter along this curve, starting with 240 ns at the lower right end and increasing until 450 ns at the upper left end. The detailed

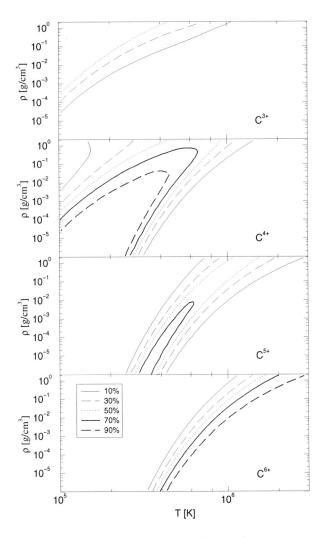

Figure 3.3: Abundances of the ions C^{6+}, C^{5+}, C^{4+} and C^{3+} given in percent of all carbon ions in the mass-density temperature plane.

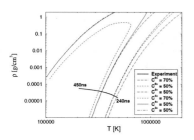

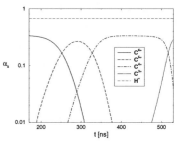

Figure 3.4: Calculated abundances of carbon ions in the mass-density-temperature plane and observed path of polyethylene plasma.

Figure 3.5: Detailed composition $\alpha_z = n_z/n$ calculated along the observed path of polyethylene plasma.

plasma composition along the evolution path of the capillary discharge plasma is given in Figure 3.5 [FÖRSTER ET AL. 98].

The charge-state distribution (Figure 3.4) is dominated at high temperatures by fully ionized carbon ions (lower right corner), followed by a narrow strip dominated by C^{5+} and a relatively large area of almost pure C^{4+} (lower left area) at medium temperatures. The shape of the different areas is mainly determined by temperature ionization and reflects the jump of ionization energies between the carbon atom and ions C^{1+} to C^{3+} ($2s$ and $2p$ electrons) on the one hand and C^{4+} and C^{5+} ($1s$ electrons) on the other hand, cf. Table 3.1.

Table 3.2: Ionization energies of the different carbon ions in RYDBERG [MOORE 93].

ion	C^{5+}	C^{4+}	C^{3+}	C^{2+}	C^{1+}	atom
ionization energy [Ry]	36.0	28.8	4.74	3.52	1.79	0.83

The effects of density ionization may be studied in the upper part of Figure 3.4 but they are relatively small in the area of the recorded path of the capillary plasma.

Discussion: In [MORGAN ET AL. 94] it was argued that fully ionized carbon needs a temperature of $\geq 160\,\text{eV}/k_B$ to appear. The characteristic ionization temperature is naturally a function of the density. According to our calculation the mass density of the discharge plasma is between 10^{-4} and $10^{-5}\,\mathrm{g/cm^3}$. In this density range a temperature of $\geq 30\,\mathrm{eV}/k_B$ is sufficient to produce a substantial amount of fully ionized carbon. Consequently, the calculation of the detailed composition (Figure 3.5) starts in the area with fully ionized carbon and continues over C^{5+} and C^{4+} to C^{3+}. This way the model considered here supports the experimental facts of the appearance of highly charged carbon in a transitory discharge regime.

3.2 Adiabatic Equation of State and Ionization Equilibrium

Adiabatic processes are thermodynamic processes which are typically characterized by a relatively fast change of state so that the system undergoing the change does not exchange heat with its surroundings. Reversible adiabatic processes are also isentropic, i.e. they take place with no change in entropy. Adiabatic processes play an important role in plasma physics and astrophysics as well as in technology.

The spectra of examples includes sound waves, expansion of plasma or gas into vacuum, ballistic-compressors, weak shocks and molecular-beam technology [FORTOV & IAKUBOV 90]. As a consequence of hydrodynamic stability the radial structure of the stars as well as of fluid planets like Jupiter or Saturn follows an adiabat [HUBBARD 89]. Finally the three adiabatic coefficients are required for stellar evolution calculations [KIPPENHAHN & WEIGERT 89] as well as in planetary and helioseismology [CHRISTIAN-DALSGAARD & DÄPPEN 92].

For many applications of an adiabatic equation of state the ideal-gas model is insufficient. One has to care for the ionization and dissociation equilibrium which determines the number of free electrons, ions of different charges, atoms, and molecules in partially ionized plasma. Furthermore, at high densities and/or low temperatures the COULOMB interaction between the charged particles and quantum effects yield contributions to, e.g., the pressure and the entropy and modifies this way the equation of state. There is a strong interplay between the interaction effects on one hand and the ionization equilibrium on the other. The more the ionization is developed and charged particles dominate the plasma composition the stronger are the COULOMB effects in the plasma. On the other hand the interaction shifts the ionization equilibrium to higher degrees of ionization, cf. section 2.3.

The aim of this section is to construct an equation of state which is given in dependence of the entropy instead of the temperature and study the behavior of the ionization equilibrium and of the equation of state along the adiabates or isentropes. First the influence of ionization on the ideal plasma isentropes is considered. Then COULOMB forces are introduced in the so called $\Lambda/8$ approximation that still allows some analytical treatment. Within this approximation the interference of the direct change of isentropes due to interaction and the indirect change through the change of ionization equilibrium is investigated. Finally an adiabatic equation of state on the basis of a PADÉ approximation is given and the adiabatic temperature gradient is obtained within this framework.

Ideal Isentropes in Partially Ionized Plasma: Consider an ideal model plasma containing only ions i, non-degenerate electrons e, and atoms a with one bound state $(a \leftrightarrow i + e)$. The reduced entropy $s = S/(V n k_B)$ with the heavy particle density $n = n_a + n_i$ is given by

$$s_{id}(T,n,\alpha) = \frac{5}{2}(1+\alpha) - \alpha \ln(\alpha y_e) - \alpha \ln(\alpha y_i) - (1-\alpha) \ln((1-\alpha) y_a) \,, \quad (3.4)$$

where $y_\nu(n,T) = n\Lambda_\nu^3/g_\nu$ with $\nu = e,i,a$ labeling the different particles species electrons, ions, and atoms and $\alpha = n_e/n$ denotes the degree of ionization. Obviously the entropy (3.4) as well as pressure $p(T,n) = (1+\alpha)n k_B T$ and thus the adiabatic equation of state $p(s,n)$ depend upon the degree of ionization. The SAHA equation (2.8) and atomic partition function for the model plasma are given by

$$\frac{1-\alpha}{\alpha^2} = n\Lambda_e^3 \cdot \sigma(T) \,, \qquad \sigma(T) = \exp\left(\frac{I}{k_B T}\right) \,, \quad (3.5)$$

where I denotes the ionization energy.

Consider equation (3.4) for the limiting cases of complete ionization $\alpha = 1$ and no ionization $\alpha = 0$. The difference in the densities belonging to the same reduced entropy s_{id} but different ionization can be obtained by eliminating temperature

$$\ln(n)\,|_{\alpha=1} - \ln(n)\,|_{\alpha=0} = \frac{s_{id}}{2} - \frac{3}{2}\ln\left(\frac{m_a}{\sqrt{m_i m_e}}\right) - \ln\frac{g_i g_e}{g_a} \approx \frac{s_{id}}{2} - 6.33 \,. \quad (3.6)$$

This means that an isentrope experiences a parallel shift in the density-temperature plane while it crosses the area of partial ionization. For a wide range of reduced entropies the adiabates start in a thin (almost ideal) atomic gas cross an area of partial ionization where interaction effects go through a maximum and return the weakly coupled region for high temperatures. The well know adiabatic equation of state for ideal gases[1]

$$pV^\gamma = \text{const.} \quad \text{and} \quad T^\gamma p^{1-\gamma} = \text{const.} \quad \text{and} \quad TV^{\gamma-1} = \text{const.} \quad (3.7)$$

where $\gamma = c_p/c_V$ applies only if the degree of ionization, i.e. the composition is constant. Equations (3.7) still hold for an ideal fermion gas and the corresponding isentropes are lines of equal degeneracy [ELIEZER ET AL. 86]. However, the specific heat ratio c_p/c_V of the fermion gas is not equal to the factor $5/3$ in (3.7) and the $c_p - c_V$ is no longer equal to R.

Adiabatic Temperature Gradient: The adiabatic temperature gradient is a particularly interesting thermodynamic quantity, e.g. it forms the basis of the SCHWARZSCHILD criterion for convective instability in stars [KIPPENHAHN & WEIGERT 89]. It is defined by

$$\nabla_{ad} = \left(\frac{d\ln T}{d\ln p}\right)_{ad} = \frac{p}{T\rho c_p} , \quad (3.8)$$

where $\delta = -(\partial \ln\rho/\partial \ln T)_p$. For an ideal model plasma with fixed degree of ionization one finds $\nabla_{ad} = 0.4$, when radiation pressure can be neglected. Therefore the adiabatic equation of state for such a system is given by a set of parallel straight lines in a $log(p)$ over $log(T)$ plot. When the degree of ionization α changes along the adiabat the generalized equation

$$\nabla_{ad} = \frac{2 + \alpha(1-\alpha)\phi_H}{5 + \alpha(1-\alpha)\phi_H^2} \quad \text{with} \quad \phi_H = \frac{5}{2} + \frac{I}{k_B T} , \quad (3.9)$$

applies cf. [KIPPENHAHN & WEIGERT 89].

Interaction Effects in the $\Lambda/8$ Approximation: The $\Lambda/8$ theory starts with an approximate expression for the chemical potential [EBELING ET AL. 76]

$$\mu = \mu_i + \mu_e = \mu_{id} - \frac{e^2\kappa}{4\pi\varepsilon_0}\left[1 - \frac{\Lambda}{8}\kappa + \dots\right] , \quad (3.10)$$

---

[1]The focus here is on low temperature plasmas and therefore radation pressure is neglected.

where κ denotes the inverse DEBYE radius and the thermal DE BROGLIE wavelength Λ of a particle with reduced mass $m = \frac{m_e \cdot m_i}{m_e + m_i}$. In the limit $T \gg I/k_B$ expression (3.10) is asymptotically exact up to the order $O(n)$ in the density. Now one constructs the simplest PADÉ approximation consistent with (3.10) in the form

$$\mu = \mu_{id} - \frac{e^2}{4\pi\varepsilon_0} \frac{\kappa}{1 + \frac{\kappa\Lambda}{8}} \ . \tag{3.11}$$

In this approximation quantum effects are expressed by just one characteristic length and one yields a DEBYE-HÜCKEL like formula where the effective diameter of the charged particles is given by $\Lambda/8$. The approximation is valid if the plasma is non-degenerate, i.e. $n_e \Lambda_e^3 \ll 1$ and yields an qualitatively correct picture even at medium densities. From thermodynamic consistency follows that all other thermodynamic functions also have a DEBYE-HÜCKEL like structure. Thus the free energy density in this model is given by

$$f = f_{id} - k_B T \frac{\kappa^3}{12\pi} \, \tau\left(\frac{\kappa\Lambda}{8}\right) \ , \quad \text{with} \quad \tau(x) = \frac{3}{x^3}\left[\ln(1+x) - x + \frac{x^2}{2}\right] \ . \tag{3.12}$$

This leads to an reduced entropy of

$$s = s_{id} - \frac{\kappa^3}{24\pi n}\left(\tau(x) + 2\tau'(x)\right) \ . \tag{3.13}$$

For high temperatures the degree of ionization α approaches unity and the ideal case is recovered. At low temperatures, when all charged particles are bound in atoms the interaction contribution of the $\Lambda/8$ model disappears and an ideal atomic-gas model remains. As compared to the ideal case the pressure

$$\frac{p}{nk_BT} = \frac{p_{id} + p_{int}}{nk_BT} = 1 + \alpha_{id} + \alpha_{int} - \frac{\alpha^{\frac{3}{2}}\kappa^3}{24\pi n}\phi(x) \ , \tag{3.14}$$

$$\phi(x) = \frac{3}{x^3}\left[1 + x - \frac{1}{x} - 2\ln(1+x)\right]$$

is altered in two ways [BEULE ET AL. 97] On the one hand pressure is increased by interaction effects due to an increase in the degree ionization as compared to an ideal model plasma, i.e. $\alpha_{int} > 0$. On the other hand the pressure is decreased by the direct contributions of the interaction effects $-\frac{\alpha^{3/2}\kappa^3}{24\pi n}\phi(x)$ to the pressure. For the entropy one may write

$$s = s_{id}(\alpha_{id} + \alpha_{int}) - \frac{\kappa^3}{24\pi n}\left(\tau(x) + 2\tau'(x)\right) \ , \tag{3.15}$$

where $s_{id}(\alpha)$ is given by (3.4). In the region where the $\Lambda/8$ approximation is valid, the entropy is increased due to the larger degree of ionization and reduced by the interaction term. The complex interplay between ionization and interaction contributions revealed in the above considerations makes it implossible to construct an analytical expression for the adiabatic equation of state or the adiabatic temperature gradient. A more appropriate description of dense plasma which includes all first order quantum corrections requires to go beyond the $\Lambda/8$ model.

Adiabatic EOS from PACH: A quantitative treatment of interaction effects for a broad range of densities and temperatures is given by the PACH approach introduced in chapter 2. The elementary constituents considered here are free electrons e, bare nuclei i, atoms a, and molecules m. n_e, n_i, n_a, and n_m denote the particle densities of these constituents. H^- and H_2^+ are not considered since their concentration are found to be negligible (smaller than 10^{-3}) even at low temperatures, $T < 10^4$ K [SAUMON & CHABRIER 92]. In the framework of this model the free energy is calculated from an expression with the following structure

$$f = f_e^{id} + f_i^{id} + f_e + f_i + f_{ie} + f_{hc} + f_a^{id} + f_m^{id} \qquad (3.16)$$

The ideal electron contribution f_e^{id} is calculated from Fermi-Dirac statistics (2.21), while for the ideal ion contribution f_i^{id} Boltzmann statistics (2.23) is used. The interaction contributions f_e, f_i, and f_{ie} consist of the three PADÉ approximations given in equation (2.25), (2.29) and (2.34) respectively. The contribution of neutral hard-sphere system f_{hc} is taken in the Carnahan-Starling approximation, cf. [FÖRSTER 92]. This free energy density is to be minimized with respect to the degree of ionization and the degree of dissociation. After performing this procedure on a dense grid in the density-temperature plane one can calculate different thermodynamic properties numerically.

Figure 3.6 [BEULE ET AL. 97] shows the isentropes $s = const.$ for hydrogen plasma together with lines of constant degree of ionization. In general the difference between ideal and non-ideal isentropes increases as density increases and temperature decreases but the interaction effects vanish when the ions and electrons are bound into to atoms and molecules. Therefore the ideal and non-ideal isentropes that nearly coincide at high temperatures merge again after they have crossed the area of partial ionization. Due to the decrease in particle number the adiabates become steeper as atoms and molecule formation takes place for decreasing temperature.

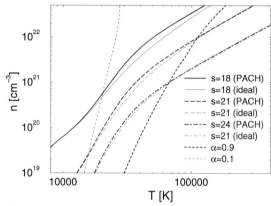

Figure 3.6: Plasma isentropes and degree of ionization in non-ideal plasma. The isentropes were obtained from PACH and in an ideal plasma model for three different values of reduced entropy $s = 18, 21, 24$. The short dashed lines limit the area where ionization raises from the weakly ionized case $\alpha = 0.1$ to the nearly completely ionized case $\alpha = 0.9$.

The degree of ionization along two of the adiabates is shown in Figure 3.7 [BEULE ET AL. 97]. Lower values of reduced entropy correspond to higher densities or lower temperatures and therefore to larger interaction effects and thus to greater differences in the degree of ionization. There are points where the interaction contributions to entropy (cf. Figure 3.6 [BEULE ET AL. 97]) cancel each other and the ideal and non-ideal isentropes cross each other. This points are found in the region of highly ionized plasma. Beyond this point the degree of ionization does not differ much for the ideal and the non-ideal case and positive contribution to the entropy due to increased degree of ionization are overcompensated by the direct interaction contributions.

The differences between ideal and non-ideal adiabatic equation of state can be seen more clearly in Figure 3.8 [BEULE ET AL. 97]. We plot the quotient of the non-ideal $p_{PACH}(s)$ and ideal pressure $p_{id}(s)$ along different adiabates over the

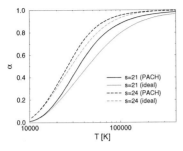

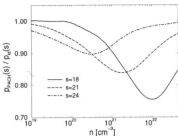

Figure 3.7: The degree of ionization along plasma isentropes $s = 21, 24$ over temperature for ideal (thin lines) and non-ideal plasma (thick lines).

Figure 3.8: Quotient of the pressure as obtained by PACH $p_{PACH}(s)$ and from an ideal model plasma $p_{id}(s)$ along isentropes for different values of reduced entropy $s = 18, 21, 24$.

density. It has to be emphasized that the pressures compared in Figure 3.8 belong to different temperatures because ideal and non-ideal adiabates take different paths through the density-temperature plane. For lower values of entropy the difference in pressure along ideal and non-ideal adiabates increase and the minimum of the quotient is shifted towards higher densities.

Once the isentropes are known it is easy to give the adiabatic (isentropic) equation of state for different values of entropy, cf. Figure 3.9 [BEULE ET AL. 97]. For high and low temperatures the ideal and non-ideal pressure almost coincide for a broad range of entropy. For intermediate temperatures interaction has significant influence on the adiabatic equation of state. The deviation of ideal and non-ideal behavior seen in Figure 3.9 for these temperatures corresponds to deviations in the degree of ionization due to interaction effects (cf. Figure 3.7). As for entropy there exist points in the area of highly ionized plasma where interaction contributions cancel each other. In order to demonstrate the effect that ionization has on plasma isentropes the adiabatic equation of state of an ideal system where no ionization processes take place is also shown. For such systems the adiabatic equation of state is given by a set of parallel straight lines in a $log(p)$ over $log(T)$ plot, cf. equation (3.8). One of this lines is plotted in Figure 3.9.

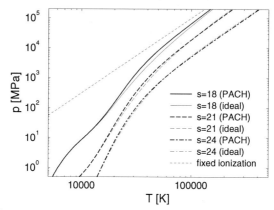

Figure 3.9: Adiabatic equation of state for reduced entropy $s = 18, 21, 24$ over temperature. We compare the pressure of an ideal plasma (thin lines) with the pressure obtained form PACH (thick lines). For comparison the adiabatic equation of state for a fixed degree of ionization as obtained from Eq. (3.8) using $\nabla_{ad} = 0.4$ is also plotted (short dashed line).

For high temperatures (i.e. a high degree of ionization) the PACH isentropes approach the ideal value, i.e. $\nabla_{ad} \approx 0.4$. But in the area where the formation of bound states takes place the adiabatic equation of state deviates significantly from this value. For decreasing temperature the pressure is decreasing much faster than in a system with fixed degree of ionization. For $s = 18$ one can even see a second steep decrease in pressure at low temperature where molecule formation sets in.

Discussion: As demonstrated within the $\Lambda/8$ model there is an interference of the direct change of isentropes due to interaction contributions and the indirect change through the shift in the ionization equilibrium. The interaction contributions to the entropy are smaller than to other thermodynamic properties like free energy or pressure. This cancellation can also be found in the limiting laws. In contrast to other thermodynamic properties, even in the case of strong coupling, entropy does not deviate much from the ideal behavior. Deviation of ideal and non-ideal adiabatic equation of state is mainly caused by different degrees of ionization along ideal and non-ideal plasma isentropes. This is due to the fact that entropy is significantly influenced by the number of particles, i.e. the degree of ionization/dissociation. As the degree of ionization is very sensitive to interaction effects the adiabatic equation of state is significantly altered in the area where bound-state formation takes place. Keeping in mind that Figure 3.6 and Figure 3.9 are plotted using a logarithmic scale one realizes that ideal and non-ideal isentropes take quite different paths through the density-temperature plane. In order to give an appropriate description of the adiabatic equation of state one has to provide a good description of the ionization equilibrium first.

4 Pressure Ionization and Plasma Phase Transition

As pointed out in chapter 1 the equation of state of dense hydrogen is still uncertain in important areas despite its central importance for astrophysics as well as for the general understanding of the behavior of matter at extreme conditions. At low temperatures and pressures, hydrogen is a molecular solid or fluid. At high pressures above 100 GPa, hydrogen is supposed to undergo an insulator-to-metal transition which has been verified experimentally for the first time in the shock-compressed fluid around 140 GPa and 3000 K by [WEIR ET AL. 96][1]. Similar conductivity data have been reported by [TERNOVOI ET AL. 99]. The physical nature of this insulator-to-metal transition at extreme conditions remains still unexplained. The interesting question, whether or not this insulator-to-metal transition is accompanied by a first-order phase transition with a corresponding instability region, a coexistence line, and a critical point have been treated within the chemical picture, see [NORMAN & STAROSTIN 70, EBELING & SÄNDIG 73, ROBNIK & KUNDT 83, EBELING & RICHERT 85b, HARONSKA ET AL. 87, SAUMON & CHABRIER 92, SCHLANGES ET AL. 95, REINHOLZ ET AL. 95, KITAMURA & ICHIMARU 98, BEULE ET AL. 99a, BEULE ET AL. 01, JURANEK ET AL. 01, JURANEK ET AL. 03]. There, the different components in a dense, partially ionized plasma such as molecules H_2, atoms H, molecular ions H_2^+ or H^-, electrons e and protons p interact via effective pair potentials. It was found that the theoretical results for the equation of state in that region depend strongly on the effective interactions for the neutral components, the molecules and atoms as well as on the COULOMB interactions between the charged particles.

---

[1]For further discussion of the experiment see [BESSON 97, NELLIS & WEIR 97].

Applying concepts of solid state theory, the observed transition can be explained by band gap closure, not accompanied by a first-order phase transition, cf. [NELLIS ET AL. 98, NELLIS ET AL. 99, NELLIS 02] for reviews. Band structure calculations which are strictly valid only for the solid at $T = 0$ K were performed for various structures. Comparing the energies of a hypothetical *bcc* metal with a close-packed lattice of hydrogen atoms or molecules interacting via effective short range potentials, the metallic structure becomes more stable than the molecular solid at high densities. Furthermore, the atomic solid has always a higher energy than the metallic solid. This picture for the solid at $T = 0$ K was adjusted by [ROSS 96, ROSS 98] to the case of a dense liquid at finite temperatures, supposing that partially dissociated hydrogen fluid is a mixture of a molecular and a metallic phase. This model includes an *ad hoc* entropy shift which is used as a fitting parameter and adjusted to represent the Hugoniots obtained from single shock experiments. Another model in the spirit of Ross was proposed in [ZINAMON & ROSENFELD 98], where the additional shift was attributed to anomalously high electronic heat and strong coupling of the electrons.

Alternatively, the behavior of dense hydrogen is derived from numerical simulations, e.g. the path-integral Monte Carlo [PIERLEONI ET AL. 96, MAGRO ET AL. 96, MILITZER & CEPERLY 00, MILITZER ET AL. 01, TRIGGER ET AL. 03, FILINOV ET AL. 03], the quantum molecular dynamics [HOHL ET AL. 93, COLLINS ET AL. 95, LENOSKY ET AL. 00, DHARMA-WARDANA & PERROT 02], or the wave-packet molecular dynamics [KLAKOW ET AL. 94a, KNAUP ET AL. 03] method, which treat a system consisting of a finite number of electrons and protons. Precursors of a first-order phase transition have also been obtained in some of these simulations [KNAUP ET AL. 03, FILINOV ET AL. 03] as well as in simulations of spin polarized hydrogen [XU & HANSEN 99] using density functional theory.

In this chapter an equation of state for dense, low-temperature hydrogen plasma is proposed by combining the methods introduced in chapter 2 for the partially ionized plasma domain with improved data for the dense, neutral fluid obtained within a dissociation model [BUNKER ET AL. 97a, BUNKER ET AL. 97b]. As before the free energy density $f = F/V$ for a two-component system of neutral (index 0) and charged particles (index $\pm$),

$$f = f_0 + f_\pm \,, \qquad\qquad (4.1)$$

is constructed. The interaction contributions in the neutral and charged subsystem
are taken into account explicitly, whereas the interaction between charges and neutrals is accounted for only by the reduced-volume concept, cf. section 2.8. Results
for the dissociation and ionization degree, the instability region, and the coexistence line are obtained for temperatures 2000 to 10000 K and densities up to
1.1 g/cm^3.

4.1   Dense Fluid Hydrogen

The equation of state for neutral, partially dissociated, fluid hydrogen can be determined applying various methods of liquid state theory such as integral equation techniques, classical Monte Carlo simulations, and fluid variational theory cf.
[BUNKER ET AL. 97a, BUNKER ET AL. 97b]. In the Monte Carlo simulations the
partially dissociated dense fluid is simulated by a mixture of H_2 molecules and H
atoms which interact via effective two-body potentials. The molecule-molecule
potential was derived by [ROSS ET AL. 83] from single shock experiments up to
pressures of 10 GPa and has been modified at low distances to avoid unphysical
behavior. The atom-atom potential is taken from [REE 88] and the atom-molecule
potential is derived by the BERTHELOT mixing rule, cf. [BUNKER ET AL. 97a, for
details]. The results for the pressure as function of the density and temperature
are in reasonable agreement with the shock-wave experiments and with the similar
linear-mixing model [ROSS 96].

Dissociation: The degree of dissociation is an input variable for the Monte-
Carlo simulations and is determined from the generalized mass action law

$$n_{H_2} = \frac{n_H^2 \Lambda_H^3}{\sqrt{2}\,\sigma_{int}} \exp\left(\frac{-D_0 + \Delta D_0}{k_B T}\right) , \qquad (4.2)$$

where the shift in dissociation energy is given by $\Delta D_0 = \mu_{H_2}^{int} - 2\mu_H^{int}$ and σ_{int} denotes the internal partition function. The interaction parts of the chemical poten-
tials where calculated from the free energies derived within the fluid variational
theory (FVT) for pure[2] molecular and atomic systems respectively using the interaction potentials mentioned before [BUNKER ET AL. 97a]. The resulting degree

---

[2]A selfconsistent treatment of dissociation energy shift and degree of dissociation was proposed
by [JURANEK & REDMER 00].

of dissociation β is shown in Figure 4.1 using the definition $\beta = n_H/n_0$, where $n_0 = n_H + 2n_{H_2}$ is the total density of neutral hydrogen while n_H and n_{H_2} are the partial densities of atoms and molecules, respectively. ρ_0, ρ_H, and ρ_{H_2} will denote the corresponding mass densities. Dissociation becomes already important in the dense fluid above 20 GPa due to the lowering of the dissociation energy with increasing density.

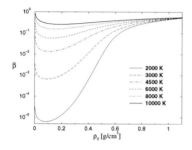

Figure 4.1: Degree of dissociation β as obtained by [BUNKER ET AL. 97a, REDMER 98a] over density for different temperatures.

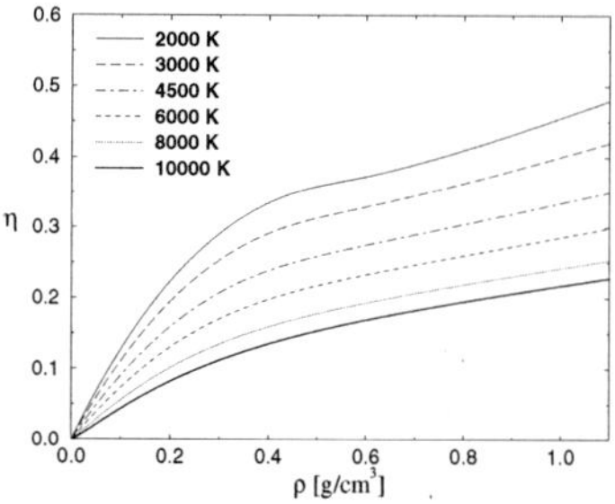

Figure 4.2: Filling parameter η as obtained from equations (4.5), using the radii given in (4.3) and (4.4) over density of fluid hydrogen ρ_o for different temperatures.

Hard-Sphere Radii: Within fluid variational theory the free energy is obtained by minimizing a free energy functional with respect to the packing fraction of a hard-sphere reference system [ROSS ET AL. 83, BUNKER ET AL. 97a]. This way density and temperature dependent hard-sphere radii are derived. For $\rho \to 0$ these density and temperature dependent radii come close to the values obtained by the prescription given by [BARKER & HENDERSON 67] and fall of allmost linear with increasing density. For the temperature range of 2000 to 10000 K and densities $\rho_{H_2} < 0.66$ g/cm^3 the hard-sphere radii as obtained by [REDMER 98a] of the H_2 molecules can be represented by

$$r_{H_2}(T, \rho_{H_2}) \approx 1.16 - 5.73 \cdot 10^{-5}\, T + 2.04 \cdot 10^{-9}\, T^2 \tag{4.3}$$
$$+ \rho_{H_2} \left(-0.486 + 2.32 \cdot 10^{-5}\, T - 4.81 \cdot 10^{-10}\, T^2 \right),$$

where the radius is in units of 10^{-10} m, the temperature T in K and the mass density ρ_{H_2} in units of g/cm^3. The density dependence of the atomic hard-sphere radii is much smaller, while the temperature dependence is similar. For the temperature range of 2000 to 10000 K and densities $\rho_H < 0.66$ g/cm^3 one may use

$$r_H(T,\rho_H) \approx 0.477 - 2.25 \cdot 10^{-5} T + 1.07 \cdot 10^{-9} T^2 \qquad (4.4)$$
$$+\rho_H \left(-4.72 \cdot 10^{-3} - 4.99 \cdot 10^{-7} T\right) ,$$

using the same units as in (4.3). From the radii formulas (4.3) and (4.4) and the degree of dissociation β one can calculate the filling parameter η of the neutral fluid

$$\eta(T,\rho_0) = \frac{4\pi}{3} \left(n_H r_H^3 + n_{H_2} r_{H_2}^3\right) , \qquad (4.5)$$

which is needed as an input quantity for the charged component of the free energy $f_\pm$. The results are shown in Figure 4.2. For low temperatures η can approach values where freezing is expected for the corresponding hard-sphere system ($\eta_{freeze} \approx 0.494$ for one-component systems [SAUMON ET AL. 89]), but it will turn out that ionization already starts at lower densities resulting in a signifincant reduction of η.

Interpolation Scheme: The pressure of the neutral subsystem p_0 can be split into the ideal contributions of the atoms and molecules and an interaction contribution,

$$p_0 = p_0^{id} + p_0^{int} = \frac{\beta\rho_0 k_B T}{m_H} + \frac{(1-\beta)\rho_0 k_B T}{2m_H} + p_0^{int}(\rho_0, T) . \qquad (4.6)$$

$\rho_0 = \rho_H + \rho_{H_2}$ denotes the mass density of the neutral particles. The data for the interaction contribution have been calculated for temperatures from 2000 to 10000 K and $\rho_0 \leq 1.1$ g/cm^3 by [REDMER 98a]. For densities and temperatures leading to low degree of dissociation the pressure agrees well with the model given by [SAUMON ET AL. 95]. These authors have used a stiffer atom-atom potential and therefore systematic deviations occur with an increasing fraction of dissociated molecules for higher temperatures and/or densities.

The Monte Carlo data can be interpolated accurately [BEULE ET AL. 99a] using an expansion of the interaction contribution with respect to the density and temperature according to

$$p_0^{int}(\rho_0, T) = \sum_{i \geq 2, j} \left(c_{ij} T^{j/2} \rho_0^i + C_{ij} T^{j/2} \rho_0^i \ln \rho_0\right) . \qquad (4.7)$$

The advantage of the expansion (4.7) is, that all other thermodynamic quantities are now at hand in a similar analytical form. For the interaction part of the internal energy one finds,

$$u_0^{int} = \sum_{i=2}^{\infty} \sum_j \frac{(1-j)}{i-1} c_{ij} \rho^i T^j + (1-j) C_{ij} \frac{1}{(i-1)^2} \rho^i T^j [(i-1)\ln(\rho)-1] . \quad (4.8)$$

An eight parameter fit with $i = 2,3,4,5$ and $j = 0,1$ is sufficient to reproduce the Monte Carlo data within typical errors of less than 4% for the pressure and less than 0.4 eV/atom for the internal energy within the region $T = (2-10) \times 10^3$ K and 0.2 g/cm$^3 \le \rho_0 \le 1.1$ g/cm^3. The respective coefficients, derived from a simultaneous least-square fit of the pressures and internal energy densities at 57 grid points in the density-temperature region given above, are listed in Table 4.1 [BEULE ET AL. 99a].

i	c_{i0}	c_{i1}	C_{i0}	C_{i1}
2	2055	-335.1	688.2	-198.6
3	0	388.2	2395	0
4	-2469	0	0	0
5	547.2	0	0	0

Table 4.1: Coefficients of the interpolation formulae (4.7) and (4.9). The pressure is obtained in GPa and the free energy density in GJ/m^3 when inserting the temperature in units of 10^4 K and the density in g/cm^3.

Free Energy Density: For the free energy density of the neutral subsystem one obtains,

$$f_0 = f_0^{id} + f_0^{int} , \quad (4.9)$$

$$f_0^{int} = \sum_{i \ge 2,j} \left\{ c_{ij} \frac{T^{j/2} \rho_0^i}{i-1} + C_{ij} \frac{T^{j/2} \rho_0^i}{(i-1)^2} [(i-1)\ln\rho_0 - 1] \right\}.$$

The ideal part depends again on the degree of dissociation via

$$f_0^{id} = n_H k_B T \left[\ln \left(\frac{n_H \Lambda_H^3}{\sigma_{PBL}} \right) - 1 \right] + n_{H_2} k_B T \left[\ln \left(\frac{n_{H_2} \Lambda_{H_2}^3}{\sigma_{H_2}} \right) - 1 \right] , \quad (4.10)$$

where the atomic and molecular partitions sums are given in section 2.3 The free energy of the charged component $f_\pm$ is treated in the same way as in chapter 2, i.e.

$$f_\pm = f_\pm^{id} + f_\pm^{int} = f_e^{id,*} + f_i^{id} + f_e^* + f_i + f_{ie} . \quad (4.11)$$

f_e^{id} and f_i^{id} denote the ideal contribution of the ions and electrons as derived from BOLTZMANN and FERMI-DIRAC statistics, respectively and the interaction contributions are given by equation (2.21) to (2.23). The asterics in (4.11) indicates that the corresponding expression is rescaled according to the concept of reduced volume for the free electrons using the filling parameter (4.5).

4.2 Plasma Phase Transition

Based on the formulae given above, the ionization equilibrium of the plasma was calculated by minimizing the total free energy. In the density region where the model shows an instability, i.e. $(\partial p/\partial V)_T > 0$, a Maxwell construction in terms of the pressure and the chemical potential was applied,

$$p(\rho_I, T) = p(\rho_{II}, T) \quad \text{and} \quad \mu(\rho_I, T) = \mu(\rho_{II}, T), \qquad (4.12)$$

where $\mu = \mu_e + \mu_i = \mu_H = \mu_{H_2}/2$ is the combined chemical potential [FÖRSTER ET AL. 92]. Figure 4.3 [BEULE ET AL. 99a] shows the results for the equation of state $p(\rho, T)$. As the pressure is primarily caused by correlation effects, the density dependence is much more pronounced than the temperature dependence.

Degree of Ionization: Within the coexistence area $\rho_I \leq \rho \leq \rho_{II}$ all intensive thermodynamic properties are constant within each phase. The combined chemical potential μ and pressure p are equal in both phases while e.g. detailed composition and mean charge $\bar{z}$ differ. Averaging over both phases is achieved by the so called lever rule, cf. [FÖRSTER 92]. For the degree of ionization one yields

$$\bar{z}(\rho) = \frac{\rho^{-1} - \rho_{II}^{-1}}{\rho_I^{-1} - \rho_{II}^{-1}} \bar{z}_I + \frac{\rho^{-1} - \rho_I^{-1}}{\rho_{II}^{-1} - \rho_I^{-1}} \bar{z}_{II} \qquad (4.13)$$

where $\rho_{I,II}$ and $\bar{z}_{I,II}$ denote the density and the degree of ionization of the coexisting phases. The results of this averaging are shown in Figure 4.4. In the partially dissociated fluid below the coexistence line the degree of ionization is always lower than 10^{-4}. Correspondingly, the conductivity is rather low, describing a semiconducting behavior. Crossing the coexistence line, the degree of ionization jumps to a value of about one third which is connected with a rather high, plasma-like conductivity, cf. section 5.5 for details.

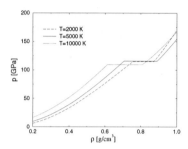
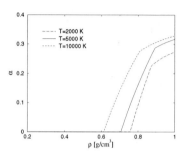

Figure 4.3: Pressure as function of the density for various temperatures. A MAXWELL construction was performed in the instability region leading to constant pressure in the coexistence region.

Figure 4.4: Degree of ionization for given temperatures versus total mass density. The kinks mark the border of the coexistence region of the PPT.

Coexistence Pressure and Coexistence Region: The phase transition which occurs in the present model is well-known as *plasma phase transition* (PPT). In the coexistence region below 10000 K, the plasma consists of two phases *I* and *II* that differ in mass densities by about 0.2 g/cm^3. The densities of the coexisting phases ρ_I and ρ_{II} have the general tendency to decline with the temperature. The PPT is characterized by a jump in the mass density, in the degree of ionization, and in the degree of dissociation. In the temperature region studied here, phase *I* is an atomic-molecular fluid with almost no ionization while phase *II* is a partially ionized plasma. Above 4000 K the transition pressure is decreasing with the temperature and varies between 120 GPa and 110 GPa for $T = (2 - 10) \times 10^3$ K. Figure 4.5 [BEULE ET AL. 99a] shows the respective coexistence line $p^{PPT}(T)$ of the two phases as well as lines of constant degree of dissociation and ionization. Figure 4.6 [BEULE ET AL. 99a] shows the density region where the two phases coexist; the coexistence region is rather narrow. The experimental point where [WEIR ET AL. 96] observed a metal-like conductivity lies above the present coexistence line supporting the result of considerable ionization in this region.

Comparison: The coexistence line of the plasma phase transition is compared with results from other approaches in Figure 4.7. Within the present

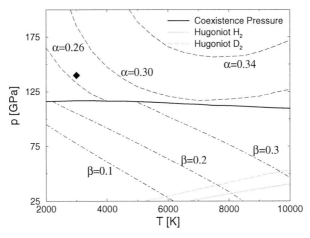

Figure 4.5: Coexistence pressure and lines of constant degree of dissociation β and ionization α, respectively, as function of the temperature. The conditions where [WEIR ET AL. 96] observed metallic conductivity is indicated by a diamond. For later reference the single shock HUGONIOT of H_2 and D_2 are also drawn (gray lines).

model the transition pressure lies in the same range as previous results obtained within the chemical [EBELING & RICHERT 85b, SAUMON & CHABRIER 92, SCHLANGES ET AL. 95, REINHOLZ ET AL. 95]. For higher temperatures $T \geq 10^4$ K in the plasma state, the location of the plasma phase transition has been studied extensively. The corresponding coexistence lines usually end in a critical point around $T_c \approx (15 \ldots 17) \times 10^3$ K, a pressure of about $p_c \approx (50 \ldots 70)$ GPa, and a density of about $\rho_c \approx (0.1 \ldots 0.5)$ g/cm^3. At much lower temperatures in the solid phase, a transition from the molecular crystal to hydrogen metal is commonly expected, cf. [MAO & HEMLEY 94]. Static high pressure experiments using diamond anvil cells allow to reach pressures of at least 340 GPa [NARAYANA ET AL. 98] at room temperature, but solid metallic hydrogen has not yet been observed using these methods. For higher temperatures $T \geq 10^4$ K in the plasma state, the location of the PPT has been studied extensively. The coexistence line ends in a critical point usually around $T_c \approx (15 - 17) \times 10^3$ K with a pressure of about $p_c \approx (50 - 70)$ GPa, and a density of about $\rho_c \approx (0.1 - 0.5)$ g/cm^3 (Figure 4.7 cf.

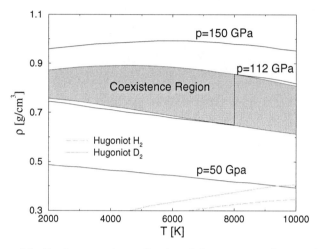

Figure 4.6: Coexistence region as function of the temperature (gray area) and three isobars: $p = 50$ GPa, $p = 112$ GPa and $p = 150$ GPa. For later reference the single shock HUGONIOT of H_2 and D_2 are also drawn (gray lines).

[BEULE ET AL. 99b]). Extrapolation of the present model towards higher temperatures matches with these predictions.

4.3   Discussion

In this chapter an equation of state of dense hydrogen plasma for $T \leq 10^4$ K was proposed giving the correct behavior for low (neutral molecular fluid) and high temperatures (highly ionized plasma).   Dissociation and ionization are treated in a consistent way by construction a free energy functional. Compared with previous approaches [EBELING & RICHERT 85b, SAUMON & CHABRIER 92, SCHLANGES ET AL. 95, REINHOLZ ET AL. 95], the correlations in the dense, neutral component are considered in a systematic way, going beyond approximative methods such as the hard-sphere reference system or perturbation theory. The applicability of this dissociation model for not too high temperatures and densities has been demonstrated by comparing the respective proton-proton distribution

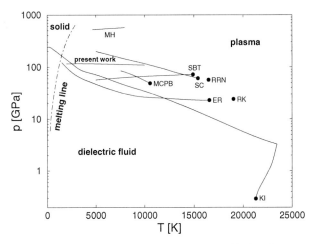

Figure 4.7: Coexistence lines (solid lines) and critical points • of the plasma phase transition given in the literature compared with the present work: [ROBNIK & KUNDT 83] labeled RK, [EBELING & RICHERT 85b] labeled RK, [MARLEY & HUBBARD 88] labeled MH, [SAUMON & CHABRIER 92] labeld SC, [SCHLANGES ET AL. 95] labeled SBT, [REINHOLZ ET AL. 95] labeled RRN, [MAGRO ET AL. 96] labeled MCPB and [KITAMURA & ICHIMARU 98] labeled. The melting line (dashed line) was taken from [ROSS ET AL. 83].

functions with results from computer simulations [NAGEL ET AL. 98].

The present equation of state yields a first-order phase transition with a corresponding separation between two phases of different mass density and degree of ionization. The coexistence pressure is comparable to previous calculations within the chemical [EBELING & RICHERT 85b, SAUMON & CHABRIER 92, SCHLANGES ET AL. 95, REINHOLZ ET AL. 95]. Numerical simulations [KNAUP ET AL. 03, FILINOV ET AL. 03] also support the occurrence of a first-order phase transition in hydrogen. Such a behavior would be very similar to that of expanded fluid alkali metals where an electronic metal-insulator transition occurs near the critical point of the ordinary liquid-vapor phase transition [HENSEL & EDWARDS 96], see [EBELING & NORMMAN 03] for discussion. In order to determine the exact location of the critical point, the present model has to be extended to higher temperatures, i.e. Monte Carlo simulations for higher temperatures are required and an explicit treatment of the charged-neutral interaction has to be considered.

The plasma phase transition is closely related to the experimentally observed insulator-to-metal transition in fluid hydrogen around temperatures of 3000 K and pressures of 140 GPa. There, a continuous electronic transition from semiconducting to metall-like conductivity occurs within a multiple shock compression experiment. Furthermore single shock experiments allow to meassure the equation of state at high pressure. The next chapter will discuss these shock wave experiments and compare equation of state and conductivity obtained within the model presented here with the available experimental data.

5 Shock Waves in Dense Plasma

A smooth density disturbance in gases or condensed matter can evolve into discontinuous quantities as a result of the nonlinear character of the conservation laws for mass, energy, and momentum. This effect had already been discovered by RIEMANN, RANKINE, and HUGONIOT in the second half of the nineteenth century, cf. [ELIEZER ET AL. 86]. Shock waves are encountered in a variety of technical problems such as supersonic flight or explosions. The study of shocks in gases and condensed matter is also of great scientific interest, as it allows to investigate thermodynamic properties of materials at high temperature and high pressure, see [ZELDOVICH & RAIZER 66, MCQUEEN 91] for reviews.

Various experiments have probed the effect of strong shocks on liquid hydrogen and deuterium cf. [HOLMES ET AL. 95, WEIR ET AL. 96, DA SILVA ET AL. 97, COLLINS ET AL. 98, MOSTOVYCH ET AL. 00, KNUDSON ET AL. 01]. The insulator-to-metal transition has been verified experimentally for the first time in a multiple shock [WEIR ET AL. 96] around 140 GPa and 3000 K. Single shock experiment with deuterium [DA SILVA ET AL. 97, COLLINS ET AL. 98, KNUDSON ET AL. 01] managed to reach pressures up to 340 GPa, but the results from laser driven shocks [COLLINS ET AL. 98] differ from those using z-pinch [KNUDSON ET AL. 01]. The pressure observed in laser driven experiment shows substantial deviation from the equation of state of the SESAME library [KERLEY 80] and therefore much interest in the theoretical descriptions of thermodynamic properties of hydrogen and deuterium at high density has been created [LENOSKY ET AL. 97, ROGERS & YOUNG 97, BUNKER ET AL. 97a, ROSS 98, BEULE ET AL. 99a, MILITZER & CEPERLY 00, BEULE ET AL. 01, JURANEK ET AL. 01, DHARMA-WARDANA & PERROT 02, JURANEK ET AL. 02,

MILITZER 03, KNAUP ET AL. 03]. The importance of these properties for various fields of physics has already been discussed in chapter 1.

This chapter first gives an overview of shock wave theory, then the effects of ionization, dissociation, and interaction on shock compression are discussed. The HUGONIOT curve for deuterium is calculated within the chemical picture and compared with experimental data and theoretical results obtained within other models. Finally multiple shocks are considered within the model proposed in the previous chapter and the corresponding results for conductivity are compared with experimental data.

The development in the next section is based upon the assumption, that the material being studied is subjected to a plane shock front and that any subsequent motion is also of that nature. Moreover, it is assumed that the material is free from any anisotropic stresses and that initial and final state are in thermodynamic equilibrium.

5.1   Hugoniot Relation

In shock wave experiments an initially homogeneous medium in a piston is compressed by a pusher moving with a velocity larger than the sound velocity[1] within the uncompressed medium cf. Figure 5.1. A shock wave that separates the already compressed from the still uncompressed part of the medium emerges and propagates through the medium.

Given the initial density ρ_0 and pressure p_0 of the uncompressed medium it is possible to determine the density ρ_1 and the pressure p_1 behind the shock front by measuring the velocity of the pusher u_p and the shock wave u_s speed and using the conservation laws for mass, energy, and momentum in the co-moving frame where the shock front is at rest [ZELDOVICH & RAIZER 66].

Conservation Laws: Let u_0 and u_1 denote the mass velocities before and after the front in the co-moving frame (Figure 5.2). The conservation of mass requires

$$j = \rho_0 u_0 = \rho_1 u_1 \,. \tag{5.1}$$

---

[1]Sound velocity $c = \sqrt{\gamma_{ad} k_B T / m}$, where $\gamma_{ad} = c_p/c_v$ denotes the adiabatic exponent.

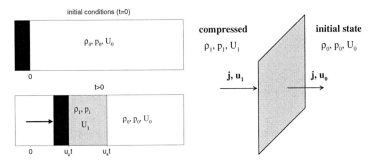

Figure 5.1: Shock wave propagation in a piston in the laboratory frame.

Figure 5.2: Shock wave in co-moving frame: $u_0 = -u_s$ and $u_1 = u_s - u_p$.

Considering the force F on the shock front area A for an infinitesimal time dt and using momentum conservation one yields

$$F\,dt = A(p_1 - p_0)dt = m_0 u_0 - m_1 u_1 \ . \tag{5.2}$$

Here $m_i = \rho_i u_i A dt$ denotes the mass current through the shock front, thus one gets

$$p_0 + \rho_0 u_0^2 = p_1 + \rho_1 u_1^2 \ . \tag{5.3}$$

Finally conservation of energy U leads to the BERNOULLI equation if dissipation can be neglected during the shock wave propagation. For a unit mass this means

$$U_0 + \frac{p_0}{\rho_0} + \frac{u_0^2}{2} = U_1 + \frac{p_1}{\rho_1} + \frac{u_1^2}{2} \ . \tag{5.4}$$

Equation (5.1), (5.3), and (5.4) can also be obtained more rigorously from fluid dynamics cf. [ELIEZER ET AL. 86, MCQUEEN 91].

Hugoniots: By introducing the specific volumes $V_i = 1/\rho_i$ and eliminating the mass current j from the conservation laws one derives the HUGONIOT relation:

$$U_0 - U_1 + \frac{1}{2}(V_0 - V_1)(p_0 + p_1) \equiv 0 \ . \tag{5.5}$$

The thermodynamic relation $U(V, p)$ is an equation of state of the material under consideration. Assuming this relation is known one gets from (5.5) a pressure versus volume (or density) curve

$$p_1 = p_H(V_1; p_0, V_0) \ , \tag{5.6}$$

that depends parametrically on the initial conditions p_0 and V_0 and is usually called HUGONIOT curve or simply HUGONIOT. The HUGONIOT is no thermodynamic path but the locus of states generated by single shocks of varying strength starting from the same initial conditions. Shock compression involves an increase in entropy, and the pressure along a HUGONIOT will lie above the isentrope starting from the same initial conditions. Because any increase in entropy is associated with a rise in temperature, the HUGONIOT also lies well above the corresponding isotherm.

Weak Shocks: In the limit of weak shock waves, defined by $p_1 \approx p_0$ the entropy production during the shock is small and the increase in pressure along the HUGONIOT is only slightly higher than along the corresponding isentrope. This is due to the fact that not only the slope of the HUGONIOT is the same as that of the isentrope, but their next derivatives are also the same [LANDAU & LIFSCHITZ 91a]

$$\left[\frac{\mathrm{d}p}{\mathrm{d}V}\right]_H = \left[\frac{\mathrm{d}p}{\mathrm{d}V}\right]_S , \qquad \left[\frac{\mathrm{d}^2 p}{\mathrm{d}V^2}\right]_H = \left[\frac{\mathrm{d}^2 p}{\mathrm{d}V^2}\right]_S . \tag{5.7}$$

For strong shocks defined by $p_1 \gg p_0$ there is a significant increase in entropy and the HUGONIOT is steeper than the corresponding isentrope [ELIEZER ET AL. 86].

Measurement and Modeling: The measurement of shock wave and pusher velocities in the laboratory frame ($u_0 = -u_s$ and $u_1 = u_s - u_p$) allows to determine the compression

$$K = \frac{V_0}{V_1} = \frac{\rho_1}{\rho_0} = \frac{u_s}{u_s - u_p} . \tag{5.8}$$

Using the momentum conservation (5.3) relation the pressure behind the shock front

$$p_1 - p_0 = \rho_0 u_0^2 - \rho_1 u_1^2 = \rho_0 u_s u_p \tag{5.9}$$

can be obtained without any assumptions about the equation of state.

Theoretical description of the equation of state are usually defined in terms of density and temperature. Given a specific model equations $U(T,V)$ and $p(T,V)$ and a set of initial conditions V_0, p_0, and U_0 one can calculate the HUGONIOT function

$$H(T,V) = U_0 - U_1(T,V) + \frac{1}{2}(V_0 - V)(p_0 + p_1(T,V)) , \tag{5.10}$$

is determined. The (numerical) solution of $H(T,V) = 0$ yields a parametric description of the HUGONIOT in terms temperatures T_1 and volumes V_1.

5.2   Compression Asymptotics

Before comparing experimental data and theoretical HUGONIOT curves it is useful
to study the influence of dissociation, ionization, and interaction on the compression asymptotics. Consider the shock wave relations for an ideal gas with constant
specific heat. It is convenient to write the equation of state as

$$U = \frac{pV}{\gamma - 1} \, . \tag{5.11}$$

By substituting (5.11) into the HUGONIOT relation (5.5) one yields

$$\frac{p_1}{p_0} = \frac{(\gamma+1)V_0 - (\gamma-1)V_1}{(\gamma+1)V_1 - (\gamma-1)V_0} \quad \text{or} \quad \frac{V_1}{V_0} = \frac{(\gamma-1)p_1 + (\gamma+1)p_0}{(\gamma+1)p_1 + (\gamma-1)p_0} \, . \tag{5.12}$$

From this equation it is clear that the compression ratio $K = V_0/V_1 = \rho_1/\rho_0$ can
not exceed a certain value determined by γ. For strong shocks $p_1 \gg p_0$ one gets

$$K_{max} = \left(\frac{V_0}{V_1}\right)_{min} = \left(\frac{\rho_1}{\rho_0}\right)_{max} = \frac{\gamma+1}{\gamma-1} \, . \tag{5.13}$$

The maximum compression that can be achieved in a ideal mono-atomic gas
$\gamma = 5/3$ is $K_{max} = 4$, while diatomic gases can be compressed by a factor of 6
if the vibrational modes are frozen and by a factor of 8 if these modes can be excited. Because the compression is always accompanied by heating, increasingly
strong shocks will eventually create temperatures where dissociation and ionization become important. In the limit of extremly strong shocks a high temperature
plasma will be generated. The compressions obtainable in this limit are discussed
next.

Dissociation and Ionization: Although it is not possible to give an analytic
solution of the coupled problem of dissociation/ionization equilibrium[2] and shock
compression one can deduce that dissociation and ionization lead to a further increase in the density ratio [ZELDOVICH & RAIZER 66]. This ratio is affected only
by that part of the specific heat that corresponds to the potential and internal energy of the particles. The increase in specific heat due to the increase in the number
of particles caused by dissociation or ionization does not directly affect the adiabatic exponent which determines the maximum density ratio K_{max}. This is easily

---

[2]For an ideal model plasma it is still possible to give a parametrization of the HUGONIOT in terms
of the final state temperature.

seen by writing the internal energy as a sum $U = U_{kin} + U_{bind}$, where U_{bind} includes the binding energy as well as the energy of the internal degrees of freedom. Substituting $p = \frac{2}{3}\rho U_{kin}$ into the HUGONIOT relation (5.5) by neglecting initial pressure p_0 (strong shock) and fixing energy scale by setting $U_0 = 0$ one yields [ZELDOVICH & RAIZER 66]

$$K = 4 + \frac{3U_{bind,1}}{U_{kin,1}} \,. \tag{5.14}$$

Using this normalization $U_{bind,1}$ is the difference in binding and internal energies between initial and final state. The difference is in general positive and therefore higher compression can be achieved $K > 4$. Note that after complete dissociation and ionization $U_{bind,1}$ remains constant while $U_{kin,1} \sim T_1$ increases further as temperature T_1 raises for stronger shocks. Therefore the ideal gas value is approached again as $T_1 \to \infty$. For initially molecular deuterium D_2 one finds from (5.14)

$$K = 4 + \frac{184370\,\mathrm{K}}{T_1} \,, \tag{5.15}$$

if the final state is assumed to be a completely ionized ideal plasma, i.e. $U_{bind,1} = I + D/2 \approx 15.88\,\mathrm{eV}$ and $U_{kin,1} = 3k_B T_1$ per atom.

Interaction Effects: Interaction can be discussed qualitatively by using the similar arguments as for ionization and dissociation, at least for non-degenerate systems. For simplicity lets assume for a moment that no dissociation or ionization processes take place during the compression. Interaction effects will give a contribution U_{int} to the energy $U = U_{kin} + U_{int}$ and additionally there is now also an interaction contribution to the pressure $p = p_{kin} + p_{int}$, but still $p_{kin} = \frac{2}{3}\rho U_{kin}$. From the fact that the interaction contributions to thermodynamic functions of classical systems depend only upon a single coupling parameter Γ [ELIEZER ET AL. 86] it follows immediately that $p_{int} = U_{int}/(3V)$. By substituting this equation into the HUGONIOT relation (5.5) and solving for the compression one gets by neglecting p_0

$$K = 4\,\frac{U_{kin,1} + U_{int,1}}{U_{kin,1} + \frac{1}{2}U_{int,1}} \,. \tag{5.16}$$

If the final state is a partial ionized plasma the interaction contribution $U_{int,1}$ and $p_{int,1}$ will in general be negative and thus K is smaller than 4, i.e. screening effects reduce the accessible compression range. Again the ideal value is approached only as T_1 goes to infinity.

The above considerations for dissociation/ionization and interaction effects can be combined. The resulting compression [BEULE ET AL. 01]

$$K = 4 \frac{U_{kin,1} + U_{int,1}}{U_{kin,1} + \frac{1}{2}U_{int,1}} + \frac{3U_{bind,1}}{U_{kin,1} + \frac{1}{2}U_{int,1}} \, , \tag{5.17}$$

shows that the effects tend to compensate each other and that dissociation/ionization will prevail for weakly non-ideal partially ionized/dissociated systems. The separation between binding energy U_{bind} and interaction (screening) energy U_{int} is well defined within the chemical picture. When using equation (5.17) for partially ionized plasmas one has to keep in mind, that interaction will also affect $U_{bind,1}$ by changing the dissociation and ionization equilibrium.

5.3   Shock Waves and Phase Transitions

The most likely type of transition to be observed in shock experiments is a phase change with an increase in density. Theory and experimental findings (mainly in solids) have been reviewed by [MCQUEEN 91]. Due to the thermodynamic constraints discussed in section 5.1 the HUGONIOT will not coincide with an isotherm, where the phase change manifest as an isobar in the (p,V) plane. In addition non-equilibrium conditions can be encountered as the phase change might be quite slow relative to the time scale of the shock wave experiment. A first order phase change is characterized by the slope of its coexistence line, which is given by CLAUSIUS-CLAPEYRON relation

$$\frac{\mathrm{d}p}{\mathrm{d}T} = \frac{\Delta S}{\Delta V} \, , \tag{5.18}$$

where ΔS and ΔV denote the difference in entropy and volume between the phases. In different models of the plasma phase transition positive [ROBNIK & KUNDT 83, HARONSKA ET AL. 87, MARLEY & HUBBARD 88] as well as negative slopes [EBELING & RICHERT 85b, SAUMON & CHABRIER 92, MAGRO ET AL. 96] of the coexistence line are found. From a phenomenological point of view the negative slope $\mathrm{d}p/\mathrm{d}T < 0$ seems to be more realistic due to an entropy production during the crossing of the coexistence area [EBELING ET AL. 88]. The resulting situation is sketched in Figures 5.3 and 5.4.

Figure 5.3 shows an equilibrium HUGONIOT in the (p,V) plane (solid line) passing through the area of a phase transition. The beginning and the end of the

coexistence area between the low (I) and high (II) density phase are marked by
arrows and manifest in kinks in the equilibrium HUGONIOT. The RAYLEIGH line[3]
connecting the initial state (p_0, V_0) with the onset of the phase transition A is shown
as a dot-dashed line. If the pressure behind the front is higher than the intersection
between the RAYLEIGH line and the HUGONIOT in the dense phase B the material
will undergo the transition within a single shock front. Otherwise one may en-
counter a double shock wave structure as a single shock front becomes unstable if
$d^2 p/dV^2 < 0$ [ZELDOVICH & RAIZER 66]. If the phase change is slow relative to
the time scale of the shock wave experiment substantial deviations from the equilibrium HUGONIOT can be encountered, e.g. at the onset the phase transition A the
transformation may not start immediately but rather follow the overdriven HUGO-
NIOT of phase I (dashed line). Furthermore this non-equilibrium situation can also
be connected with a substantial increase of the front width, cf. [MCQUEEN 91].

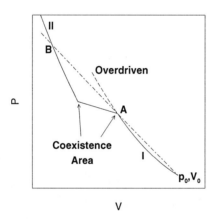

Figure 5.3: Sketch of an equilibrium HUGONIOT in the (p, V) plane (solid line)
passing through the coexistence area (arrows) between phase I and II together
with the overdriven HUGONIOT (dashed line) of phase I and a RAYLEIGH line
(dot-dashed line).

As the temperature changes along the HUGONIOT it is not possible to draw a co-

---

[3]A RAYLEIGH line is a straight line in the (p, V) that connects initial (p_0, V_0) and final state
(p_1, V_1) of a shock. The slope of the RAYLEIGH line is proportional to the shock velocity -
higher compressions are associated with faster shocks.

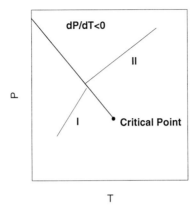

Figure 5.4: Sketch of the coexistence curve (thick solid line) and critical point of a phase transition with $dp/dT < 0$ in the (p, T) plane and an equilibrium HUGONIOT in low density phase I and high density phase II (thin solid lines).

existence area in Figure 5.3. The situation is more clear in the (p, T) plane shown in Figure 5.4. If the phase transition has a coexistence line with negative slope the HUGONIOT will eventually cross the coexistence line from the low density phase I to the high density phase II provided it does not pass below the critical point.

5.4 Single Shock Hugoniot

The considerations here focus on deuterium D_2 with the initial conditions that are typical for the single shock experiments, i.e. liquid deuterium with a density of $\rho_0 = 0.171$ g/cm^3 corresponding to $V_0 = 0.391 \cdot 10^{-25}$ cm^3 per molecule. The initial temperature in these experiments is $T_0 = 19.6$ K and the initial pressure is below 10^5 Pa. In the NOVA experiments [DA SILVA ET AL. 97, COLLINS ET AL. 98] laser-pulse generated shocks with speeds of the order $3 \cdot 10^4$ m/s allowed to reach pressures up to 340 GPa and about six-fold compression. Below the minimum pressure of the laser-pulse experiment (23 GPa) the deuterium HUGONIOT has been probed using a double stage gas-gun [NELLIS ET AL. 83, HOLMES ET AL. 95]. While the final state temperature is not available in the

Figure 5.5: (next page) Pressure p, degree of dissociation β, and temperature T along a single shock HUGONIOT curve of D_2 versus density: diamonds $\Diamond$ and dots • denote data from laser-driven shock wave experiments [DA SILVA ET AL. 97, COLLINS ET AL. 98], while points denoted by $\times$ and $+$ are taken from gas-gun experiments [NELLIS ET AL. 83, HOLMES ET AL. 95]. Squares $\square$ denote recent measurements by [KNUDSON ET AL. 01]. The present results in the low and high temperature branch (thick solid lines) are compared with HUGONIOT curves derived from the linear-mixing model (LM) [ROSS 98], the SESAME tables [KERLEY 80], tight-binding molecular dynamics (TBMD) [LENOSKY ET AL. 97] and path-integral Monte Carlo simulations (PIMC) [MILITZER & CEPERLY 00], wave packet molecular dynamics (WPMD) [KNAUP ET AL. 03], the activity expansion method (ACTEX) [ROGERS & YOUNG 97], the hydrogen equation of state (SC) given in [SAUMON & CHABRIER 92] and the modified linear mixing model of [ZINAMON & ROSENFELD 98]. [JURANEK ET AL. 01] (JRS) included polarisation effects in the high temperature PACH model and improved the FVT model for pressure dissociation [JURANEK ET AL. 02] (JRR).

NOVA experiments it was measured spectroscopically by [HOLMES ET AL. 95] in their gas-gun experiment. Another shock wave compression experiment using the z-pinch technique [KNUDSON ET AL. 01] yielded substantially lower compression (four fold) and reached pressures up to 70 GPa.

To compare the results for the equation of state presented in chapter 4 directly with available experimental data, pressure p, degree of dissociation β, and temperature T along a HUGONIOT curve of deuterium was calculated. The resulting HUGONIOT is shown in Figure 5.5 (following [BEULE ET AL. 99a]) together with the experimental data as well as other theoretical calculations. In order to use the hydrogen equation of state for deuterium mass scaling is applied for the interpolation formula of the interaction contributions (4.9), i.e. it is assumed that equal temperatures and particle numbers for hydrogen and deuterium lead to an equal degree of dissociation and equal interaction contributions to the thermodynamic function of the neutral fluid.

Low Temperature Branch: The temperatures considered in the previous chapter $(2\,000 - 10\,000$ K$)$ correspond to densities of about $0.4 - 0.8$ g/cm^3. In

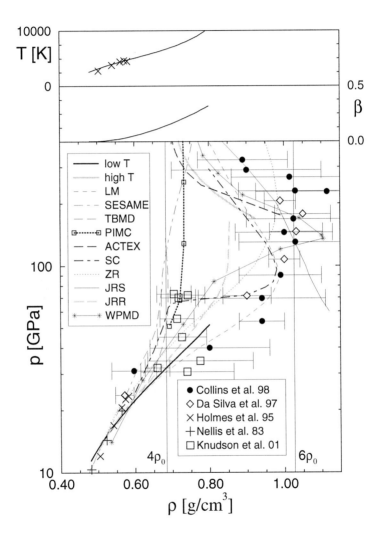

this range the proposed model (labeled low T) is in good agreement with the experimental data, the linear-mixing model (LM) of [ROSS 96, ROSS 98], and the modified linear-mixing model of [ZINAMON & ROSENFELD 98] (ZR). The calculated final state temperatures are in good agreement with the measurement of [HOLMES ET AL. 95]. The strong increase in compression is associated with an strong increase in dissociation, while the degree of ionization remains lower than 10^{-4} in the considered temperature range. Deviations from the HUGONIOT behavior predicted by the SESAME tables [KERLEY 80] and the measurements by [KNUDSON ET AL. 01] occur mainly due to this dissociation, see also [JURANEK ET AL. 01]. The tight-binding molecular dynamics simulation (TBMD) of [LENOSKY ET AL. 97] as well as [LENOSKY ET AL. 00] does not reproduce the strong compression between 30 GPa and about 100 GPa but rather reproduce the behavior of the z-pinch experiments. Applying the hydrogen equation of state given by [SAUMON & CHABRIER 92] for the calculation of the HUGONIOT curve (SC), an almost abrupt compression at about 70 GPa occurs. The wave packet molecular dynamics simulations [KNAUP ET AL. 03] produce a similar but smoother transition from four fold to about six fold compression near 100 GPa. The path-integral Monte Carlo simulations (PIMC) [MILITZER & CEPERLY 00] become increasingly reliable as the temperature increases, but can not yet reproduce the neutral fluid behavior at low temperatures.

High Temperature Branch: At high pressures, i.e. high temperatures, the HUGONIOT curve enters the region of highly ionized plasma. In this area the simplified treatment of the neutral component proposed in section 2.7 is applicable. The corresponding HUGONIOT (labeled high T) shows good agreement with the activity expansion method (ACTEX) of [ROGERS & YOUNG 97] and reasonable agreement with the path integral Monte Carlo simulation [MILITZER & CEPERLY 00], the linear mixing model [ROSS 98], with the [SAUMON & CHABRIER 92] equation of state and the results of [JURANEK ET AL. 01, JURANEK ET AL. 02]. The results of the linear mixing type models (LM and ZR), the activity expansion ACTEX, and the SC equation of state, and the PACH approach are in reasonable agreement with the experimentally observed turn-around of the HUGONIOT curve around 200 GPa. This turn-around is not reproduced by the SESAME tables or the TBMD simulation or in the PIMC simulations.

Compression Asymptotic: At about 300 GPa the majority of the theoretical HUGONIOT curves converge at a density that is slightly lower than the experimentally observed one. For even higher pressures the consideration given in section 5.2 supply strong restriction on the HUGONIOT. Figure 5.6 (following [BEULE ET AL. 00]) depicts these restrictions together with the highest experimentally obtained pressures and HUGONIOTs determined from PACH [BEULE ET AL. 01], ACTEX [ROGERS & YOUNG 97], PIMC [MILITZER & CEPERLY 00], WPMD [KNAUP ET AL. 03] and within the DEBYE approximation.

The dotted line (labeled binding) corresponds to the HUGONIOT of an ideal plasma model that neglects all interaction effects but includes the difference in binding energy between initial and final state. In the area of complete ionization a parametric representation of the HUGONIOT is given by equation (5.15). The short dashed curve (labeled interaction) corresponds to a model plasma that treats the interaction within DEBYE approximation but ignores the difference in binding energy between initial and final state, cf. equation (5.16). When considering the effects of both interaction and binding energy within the DEBYE model they tend to compensate each other, see equation (5.17).

As predicted in section 5.2 the DEBYE HUGONIOT approaches compression $K = 4$ from above. Beyond 30 000 GPa the ideal plasma model is sufficient. Below this point interaction effects lead to a reduced compression. The ACTEX and PACH curves separate at about 4000 GPa corresponding to a temperature[4] of about 600 000 K and a coupling parameter of $\Gamma_i \approx 0.3$ in a region of almost complete ionization. The DEBYE model and ACTEX prediction coincide down to pressures of approximately 1000 GPa and temperatures of about 200 000 K, i.e. a region where the coupling parameter is already larger than one. The PACH as well as ACTEX HUGONIOT turn sharply towards higher density as recombination sets in and therefore interaction contributions are reduced. It can be concluded that bound state formation (formation of atoms) can contribute to increased compressibility (above four-fold) just as well as the existence of dia-atomic molecules. This effect leads to the differences to the PIMC HUGONIOT at pressures below 300 GPa. For lower temperatures the DEBYE approximation severely overestimates interaction contribution and thus the corresponding HUGONIOT turns towards lower densi-

---

[4]The temperatures given in this paragraph where calculated using PACH.

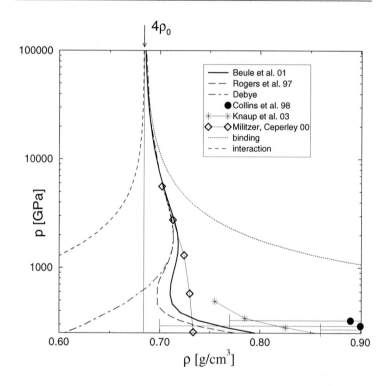

Figure 5.6: Deuterium HUGONIOTs for high pressures as obtained from PACH [BEULE ET AL. 01] (solid line), ACTEX [ROGERS & YOUNG 97] (long dashed line), PIMC [MILITZER & CEPERLY 00] (diamonds), WPMD [KNAUP ET AL. 03] (stars), and the DEBYE approximation (dashed-dotted line). The highest experimentally accessible points [COLLINS ET AL. 98] are shown for comparison. The two remaining curves illustrate the effects of interaction and binding energies on the asymptotic behavior of the compression. The dotted curve (labeled binding) corresponds to an ideal plasma model that neglects all interaction effects but includes the difference in binding energy between initial and final state. The short dashed curve (labeled interaction) corresponds to a model plasma that treats the interaction within DEBYE approximation but ignores the difference in binding energy between initial and final state.

ties. The PACH HUGONIOT passes the highest observed pressures of 340 GPa
at slightly lower densities than experimentally determined with a temperature of
75 000 K. The predicted degree of ionization for these conditions is 0.45. For such
conditions the quantitative agreement with the experiment is limited by the simplified treatment of neutral particles in the high temperature branch. The accessible
WPMD data show a similar compressibility behavior as PACH and ACTEX but at
higher pressures.

Discussion: The strong deviations between the different theoretical curves and
experiments found in Figure 5.5 have generated a fierce discussion on the physical reasons for these discrepancies.   The molecular-dynamics simulations of
[LENOSKY ET AL. 97] are based on tight binding potentials that are chosen to best
reproduce a set of ab-initio results. No evidence for a phase transition within this
region is found and the HUGONIOT is very similar to the one obtained by the
SESAME data. An ad-hoc change of the binding energy allows to deduce that an
additional energy of $3 - 5\,eV$/atom would be needed to reproduce the experimentally observed densities within this model. The authors conclude that such a large
energy shift can not be attributed to molecular dissociation and cast doubts about
thermodynamic equilibrium in the NOVA experiment. As discussed in section 5.2
the maximum possible compression for a molecular (di-atomic) and an atomic
medium differs by a factor two due to the energy that is stored in the internal degrees of freedom of the molecules. Therefore the predictions for the density along
a HUGONIOT curve in a partially dissociated molecular fluid depends sensitively
on the degree of dissociation. The role of dissociation for the HUGONIOT is also
emphasised by [ROSENFELD 01]. This effect does not depend primarily on the
magnitude of the dissociation energy but rather on the existence of rotational and
vibrational modes within the molecules. As discussed in [LENOSKY ET AL. 99]
the model used for the neutral phase in the PACH calculation reaches higher compression than the TBMD due to a later and slower release of the H_2 binding energy.
Therefore the degree of dissociation is lower when the vibrational degree of freedom becomes accessible due to temperature increase during compression. This
additional degree of freedom allows for substantially higher compression as discussed in section 5.2.

The path integral Monte-Carlo simulations utilize a decomposition of
the density matrix into a product of high temperature density matrices
[PIERLEONI ET AL. 94, MILITZER & CEPERLY 00, FILINOV ET AL. 03]. The re-

sulting path integral is evaluated using a METROPOLIS algorithm and the fermion sign problem is treated by introducing nodal restrictions. PIMC simulations become increasingly reliable as temperature increases, thus [MILITZER ET AL. 98] conclude that the discrepancy at pressures higher than 100 GPa is a fundamental problem. Calculations by [MILITZER & CEPERLY 00] show a behavior similar to [KERLEY 80], while [FILINOV ET AL. 03] reported droplet formation that can be interpreted as an indication of a first-order phase transition. The computational requirements that increase substantially with decreasing temperature confine the PIMC method to fairly small values for sample size. No real phase separation can take place in the simulations especially when using periodic boundary conditions, cf. also [FILINOV ET AL. 03]. When the plasma enters the coexistence area of the phase transition by change of density or temperature the whole simulation volume will rather stay in the dominant phase. Thus the simulation will not follow thermodynamic equilibrium situation while traversing the coexistence area. The physical reason for this behavior is that the creation of small bubbles of the other phase would require large surface energies. Due to the same reasons a possible first-order phase transition will be hard to detect in any simulation method using small simulation volumes. The same reasoning applies to the WPMD simulations [KNAUP ET AL. 03].

The linear mixing model has the attractive property of reproducing the experimental gas-gun data [ROSS 98]. It includes molecular dissociation through a mass action law and is designed to reproduce data from single and double shock experiments. The molecular fluid phase is calculated from fluid variational theory using the inter-molecule potential of [ROSS ET AL. 83]. The mono-atomic phase is approximated by a metallic fluid equation of state with an additional energy shift that serves as the fit parameter. The degree of dissociation is determined by minimizing the resulting energy functional. The free energy of the fluid mixture is determined by a composition average of the free energy of the pure molecular and metallic phase within the linear mixing approximation. Using the linear mixing model [DA SILVA ET AL. 97, COLLINS ET AL. 98] conclude that their HUGONIOT data strongly indicate a dissociative transition from the di-atomic to the mono-atomic fluid state. However, [ZINAMON & ROSENFELD 98] pointed out that an adjustable free energy shift $\Delta F = -T\delta$ in the linear mixing models appears to be unphysical, since the OCP model used for the mono-atomic state does not seem to leave room for such large errors in the free energy. These authors obtain a modified linear mixing model by accounting for the anomalous high electronic heat capacity and strong coupling of the electrons. The resulting model contains two adjustable parameters whose values are known in the ideal limit. [JURANEK ET AL. 02] ruled out the non-additivity effects as a possible explanation for the difference in experimental results and theoretical calculation.

The differences between the available experimental data [KNUDSON ET AL. 01] and [COLLINS ET AL. 98] are not yet understood [KNUDSON ET AL. 03]. [NELLIS 02] argued that [KNUDSON ET AL. 01] results are more consistent with measurements for other low-Z elements and concluded that a temperature driven nonmetal-metal transition occurs at about 50 GPa. The density functional simulations by [DHARMA-WARDANA & PERROT 02] hint non-equilibrium electron and ion temperature as a possible explanation of the differences. When using equal electron and ion temperature the HUGONIOT was quite close to the SESAME behavior, but the non-equilibrium case with ions temperature higher than electron temperature $T_i > T_e$ provided a HUGONIOT similar to the one obtained in the NOVA experiments.

Plasma Phase Transition: Below temperatures of 10 000 K the coexistence line and coexistence area of the phase transition lie subtantially above the single shock HUGONIOT curves for hydrogen and deuterium, see Figure 4.5 and 4.6. The difference between the two HUGONIOT curves stems mainly from the fact that the initial density of fluid hydrogen (0.071 g/cm^3) is lower than half the fluid deuterium density. By the time the HUGONIOTs reach the range of the predicted coexistence pressure they have generated temperatures that are in the range of the predicted critical points, i.e. the HUGONIOTs will not cross the coexistence region or cross it close to the critical point. Given the peculiar experimental situation and the large deviations of the different theoretical modeling, single shock experiments are not particular suited for probing the existence or non-existence of a first-order plasma phase transition.

5.5   Multiple Shock Experiments

Multiple shock experiments allow to generate high pressures at lower temperatures than single shock experiments. These multiple shocks can be generated by reflecting a shock wave at a boundary of high shock impedance (the product of density and shock velocity). Experiments with reflected laser-driven shocks reached pressures up to 600 GPa [MOSTOVYCH ET AL. 00]. In the experiment of [WEIR ET AL. 96] a layer of liquid H_2 or D_2 is compressed by a shock reverberating between two stiff Saphir anvils. The compression is initiated by a gas-gun accelerated impactor hitting one of the anvils. The final pressure is obtained by shock impedance matching the HUGONIOTS of the impactor and the anvil, see [ZELDOVICH & RAIZER 66, NELLIS ET AL. 83] for details. The first of the reverbarating shocks[5] is a strong shock ($p_1/p_0 \approx 50\,000$) and therefore follows the single shock HUGONIOT, however the successive reverberations are much weaker ($p_1/p_0 < 4$) and can therefore be approximated by an isentrope [ROSS 96, NELLIS ET AL. 99].

Using this approach within the linear mixing model [ROSS 96] determines a temperature of 2600 K at a density of 0.7 g/cm^3. The temperature of 3000 K given [WEIR ET AL. 96] was based on a hydrodynamics computer simulation using the SESAME equation of state. The differences are attributed to the dissociation, that reaches 6.4% at the conditions where the insulator-to-metall transition is observed [ROSS 96].

Figure 5.7 [BEULE ET AL. 01] shows isentropes for different values of reduced entropy together with the single shock HUGONIOT and the coexistence line for hydrogen within the PACH approach. According to [ROSS 96] the first shock will reach a pressure of 4.7 GPa and temperature of 1450 K. The isentrope starting at these conditions ($s \approx 7$) reaches a pressure of 140 GPa at a temperature below 2000 K and a density of about 0.9 g/cm^3. These values relie on the extrapolation of the present model below 2000 K and are therefore not very accurate. Nevertheless the lower temperature as compared to [ROSS 96] is consistent with the higher degree of dissociation and the additional ionization found in the PACH approach at these conditions. Like in single shock experiments ab initio PIMC and GGA simulation [MILITZER ET AL. 01] predict a smaller compressibility than observed experimentally.

Conductivity: The electrical conductivity σ and other transport coefficients in non-degenerate, low-density plasmas are well described by the SPITZER theory [SPITZER 57]. For strongly coupled, degenerate plasmas such as fluid metals the

---

[5]A detailed modeling of these processes comparing different equations of state was given by [TAHIR ET AL. 03].

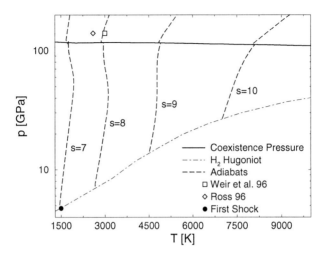

Figure 5.7: Isentropes for different values of reduced entropy *s* together with the single shock HUGONIOT and the coexistence line for hydrogen. The square □ and the diamond ◊ mark the temperature for the insulator-to-metal transition as given by [WEIR ET AL. 96] and [ROSS 96], respectively. The filled circle • marks the final state of the first shock.

ZIMAN formula is applicable. A general approach to the transport properties of COULOMB sytems valid for arbitrary coupling and degeneracy has been derived within linear response theory (see [RÖPKE 88, REINHOLZ ET AL. 95]). The model introduced in chapter 4 was used to determine the relative fraction of free electrons, their mobility is determined by the scattering processes at ions and atoms. The corresponding transport cross sections were calculated on T-matrix level within a partial wave analysis using the generalized DEBYE potential for the charged particle interactions and the polarization potential for electron-atom interaction [REDMER ET AL. 99]. A similar approach has been applied successfully to describe the transition from metallic to non-metallic conductivities that have been observed in aluminium and copper plasmas [REDMER 99] and the semiconducting behavior of hydrogen [REDMER 98b] observed in single shock experiments by [NELLIS ET AL. 92].

The principle behavior of the conductivity for fluid hydrogen as function of the

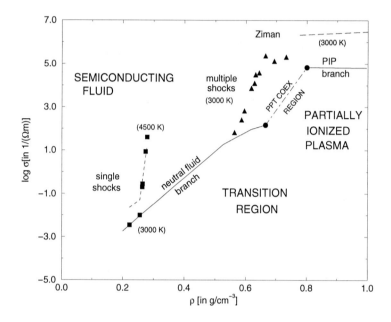

Figure 5.8: The results from single shock experiments between 3000 K and 4500 K (filled boxes [NELLIS ET AL. 92]) can be attributed to thermally activated transport in a weakly ionized plasma using a partially ionized plasma model [REDMER 98b] (dashed line). The multiple shock experiments for an almost constant temperature of 3000 K (filled triangles [WEIR ET AL. 96]) show a saturation at metall-like conductivities near 0.7 g/cm³. The 3000 K isotherm of the present model has two branches (solid lines) outside the coexistence region of the plasma phase transition which is indicated by the dash-dotted line between the filled circles. Semiconducting behavior is characteristic for the neutral fluid branch while plasma-like conductivities occur for the high-density branch [REDMER ET AL. 99]. The Ziman formula (long-dashed line) is relevant for a fully ionized plasma.

mass density is shown in Figure 5.8, cf. [REDMER 98b]. In order to reproduce
the typical semiconducting behavior in the low-density branch screening had to
be taken into account in the ionization equilibrium. The transition to metallic-like
conductivities in the multiple shock experiments [WEIR ET AL. 96] is observed
near 0.7 g/cm^3 but the limiting curve for a fully ionized plasma given by the ZI-
MAN formula is not yet reached. The overall agreement of the present model is
reasonable. The low as well as high-density branch are consistent with the experimental data but the transition to metall-like conductivities appears at slightly
higher densities than in the multiple shock experiments. Within the coexistence region the conductivity is difficult to predict as it depends on the details of the phase
segregation, but it will be dominated by the metall-like conductivity of the high
density phase. Therefore the dash-dotted line in the coexistence region may only
serve as a lower bound for the conductivity in this area. Besides thermally activated
transport, self-doping and bandgap closure have been suggested as possible mech-
anisms for the transition from typical non-metallic conductivities to metallic-like
conductivities in fluid hydrogen at Mbar pressures, see [NELLIS ET AL. 98] for a
review.

5.6 Conclusion

In this chapter the PACH approach was used to calculate the HUGONIOT curves
of deuterium with initial condition corresponding to single shock experiments
as well as the properties of hydrogen in multiple shock experiments. For high
temperatures the single shock HUGONIOT obtained within PACH is in good
agreement with the activity expansion [ROGERS & YOUNG 97] and path integral
Monte-Carlo [MILITZER & CEPERLY 00] results. Using the improved treatment
of the neutral component proposed in chapter 4 one finds good agreement with
the data from the gas-gun [NELLIS ET AL. 83, HOLMES ET AL. 95] and laser-
pulse [DA SILVA ET AL. 97, COLLINS ET AL. 98] generated single shock experiments as well as with the linear mixing model [ROSS 98]. z-pinch data by
[KNUDSON ET AL. 01] are more consistent with PIMC and SEASAME.

The behavior of the electrical conductivity as found in the multiple shock experiments [WEIR ET AL. 96] is qualitatively reproduced using this equation of state.
Compared with the model of [ROSS 96] a higher degree of dissociation and there-

fore a lower temperature is predicted for the final state in this experiment.

The calculations presented here within the PACH approach show that the large increase in density between 30 and 100 GPa observed in the single shock experiments [DA SILVA ET AL. 97, COLLINS ET AL. 98] as well as the conductivity data of [WEIR ET AL. 96] are consistent with the existence of a plasma phase transition. It seems hasty to draw final conclusions about the nature of the insulator-to-metal transition on the basis of the linear mixing model as done by [HOLMES ET AL. 95, WEIR ET AL. 96] as this model contains an energy shift as a fitting parameter that is introduced without detailed physical reasoning.

Given the peculiar experimental situation that is reflected in the large error bars on the current experimental data and the theoretical challenges of modeling dense plasmas it must be concluded that currently neither the single shock nor the multiple shock experiments allow to finally exclude or confirm the existence of a first-order plasma phase transition. Thus still refined experiments as well as theoretical modeling are required to clarify the nature of the observed insulator-to-metal transition[6]. Due to the proximity of single shock HUGONIOT and the predicted critical point multiple shock experiments seem more appropriate to clarify the situation. A temperature measurement in these experiments would be especially desirable because it would allow for the validation of theoretical models at high pressures and low temperatures.

---

[6]A similar point of view was also taken by [WEIR 98].

6 Stochastic Kinetics of Electron Transitions in Dense Plasmas

An equilibrium description may be insufficient for strongly coupled plasmas that are generated by high energy deposition on short time scales. In many situations it is still reasonable to use evolution equations for the macroscopic variables like particle densities, temperature or degree of ionization. The plasma is usually composed of molecules, atoms, and ions in different charge and excitation states, as well as of free electrons, which undergo mutual transformations by ionization and recombination, excitation and de-excitation, dissociation and association, or by combinations of these elementary processes, cf. [BIBERMAN ET AL. 87, CAPITELLI & BRADSLEY 90].

Non-equilibrium partially ionized plasma may develop rather complex reaction kinetics. The kinetic description of electron transitions between different bond states and of ionization and recombination processes with the continuum in terms of rate equations is a well elaborated field for ideal plasmas [GRIEM 68, BIBERMAN ET AL. 87, DRAWIN & EWARD 77, SAWADA & FUJIMOTO 95]. In dense plasmas the theory of these processes can be based on the GREENs function method [KRAEFT ET AL. 86] that describes the propagation of one or more specified particles within the many-particle system and allows to reveal static and dynamic properties in terms of multi-particle response and correlation functions. It results in generalized quantum kinetic equations [KREMP ET AL. 89, SCHLANGES & BORNATH 93] for the phase-space distribution of the different particle species. In many cases it is reasonable to integrate over the momentum distribution [DRAWIN & EWARD 77, CAPITELLI & BRADSLEY 90,

BIBERMAN ET AL. 87] and one yields a systems of coupled partial differential equations for macroscopic variables like particle densities, temperature or degree of ionization [EBELING ET AL. 89, OHDE ET AL. 95]. The interplay between ionization/recombination processes and temperature evolution is of special interest in partially ionized plasma, as these processes are connected with the consumption/production of large amounts of kinetic energy [OHDE ET AL. 95, BEULE ET AL. 96, BORNATH ET AL. 98]. For spatial homogeneous systems the partial differential equations reduce to a set of coupled ordinary differential equations called rate equation. Much work has been devoted to the proper inclusion of interaction effects into this rate equations [KREMP ET AL. 89, LEONHARDT & EBELING 93, BONITZ 90, BORNATH ET AL. 01] resulting in generalized rate coefficients that introduce additional non-linearities. For inhomogeneous systems the interaction effects also affect the transport coeffiecents, [OHDE ET AL. 97, BORNATH ET AL. 98]. It turns out that excitation rates are exponentially enhanced and that the area of strong coupling may involve collective effects like bistability, nucleation, and front propagation [EBELING ET AL. 87, EBELING ET AL. 89, KREMP ET AL. 89, BONITZ 90, FÖRSTER 92, SCHLANGES & BORNATH 93].

In this chapter a replacement for the deterministic differential equations based on a master equation is proposed and applied to homogeneous as well as inhomogeneous systems. The advantage of the stochastic description is twofold: First it allows for efficient simulation techniques and second the effects of fluctuations can be studied. The fluctuations will become important for weakly occupied states, systems with a mesoscopic number of charges, e.g. dusty plasmas, and in the vicinity of the critical point of the plasma phase transition. Using semi-empirical cross sections the method is applied to investigate the recombination of dense plasmas containing carbon and hydrogen. Many-body effects are taken into account via energy-level shifts that are determined by the interaction parts of the chemical potentials of free and bound particles. The master equation can be combined with a temperature equation to account for different thermodynamic constraints. Isentropic, isothermal, and isoenergetic recombination of dense hydrogen plasma are compared. For inhomogeneous systems the main focus is set on the propagation of ionization fronts with different geometries. The shape of the fronts found in the stochastic simulation compares well analytical results but non-pertubative fluctuation effects generate a diffusive instability of the front position and alter the mean velocity. Before proceeding to the stochastic description a comprehensive overview of the deterministic description is given in the first section in order to provide test cases for the stochastic approach.

6.1 Macroscopic Description

Consider an inhomogeneous reacting partially ionized dense hydrogen-like plasma. The elementary constituents are free electrons e, bare nuclei i, and atoms a. In order to obtain a model plasma which allows for a straightforward comparison between simulations and analytical results excited states and molecule formation will not be considered. The elementary reactions in this model plasma are impact ionization $a + e \longrightarrow i + e + e$ (Figure 6.1) and three-body recombination $i + e + e \longrightarrow a + e$ (Figure 6.2).

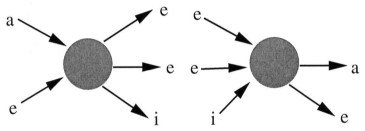

Figure 6.1: Electron-impact ionization   **Figure 6.2:** Three-body recombination

Mean-field reaction-diffusion equations for reacting dense plasma have been derived from quantum statistics [EBELING ET AL. 91, EBELING ET AL. 87, EBELING ET AL. 89, KREMP ET AL. 89, BONITZ 90, FÖRSTER 92, SCHLANGES & BORNATH 93, OHDE ET AL. 97]. Assuming homogeneous total pressure and using a suitable coordinate system one finds the following evolution equations

$$
\begin{aligned}
\frac{\partial n_e}{\partial t} + \nabla j_e &= w_e(n_e, n_i, n_a, T) \ , \\
\frac{\partial n_i}{\partial t} + \nabla j_i &= w_i(n_e, n_i, n_a, T) \ , \\
\frac{\partial n_a}{\partial t} + \nabla j_a &= w_a(n_e, n_i, n_a, T) \ ,
\end{aligned}
\tag{6.1}
$$

where $n_e(\mathbf{r},t)$, $n_i(\mathbf{r},t)$, and $n_a(\mathbf{r},t)$ denote the particle densities of electrons, ions, and atoms, respectively. $w_e = w_i = -w_a$ are the reaction source functions and $j_\nu = -D_\nu \nabla n_\nu$ ($\nu = e, i, a$) denote the currents with diffusion coefficients D_ν. Only for strong assumptions allow analytical solutions of (6.1) for relaxation times, profiles, and velocities of ionization fronts, and for the parameters of critical nucleus. Numerical solutions were achieved for plane fronts and spherically symmetrical nucleation where the equations reduce to the quasi one-dimensional case, cf. [BONITZ 90, FÖRSTER 92].

The source function describes the effects of impact ionization and three-body recombination,

$$w_e = \alpha n_a n_e - \beta n_i n_e^2 , \qquad (6.2)$$

where α and β are rate coefficients. Radiative transitions are not considered here as they do not contribute significantly above a certain density [KIPPENHAHN & WEIGERT 89, CAPITELLI & BRADSLEY 90].

Rate Coefficients: The rate coefficient for ionization of atoms embedded in an ideal thermal plasma has been derived in [BIBERMAN ET AL. 87]

$$\alpha_{id} = -\frac{10\pi a_B^2 I}{\sqrt{2\pi m_e k_B T}} \, \mathrm{Ei}\left(\frac{-I}{k_B T}\right) , \qquad (6.3)$$

where I denotes the ionization energy and a_B the Bohr radius. For $k_B T \ll I$ this rate coefficients can be approximated by

$$\alpha_{id} = \frac{10\pi a_B^2 k_B T}{\sqrt{2\pi m_e k_B T}} \, \exp\left(\frac{-I}{k_B T}\right) . \qquad (6.4)$$

Ionization and three-body recombination rate coefficients are connected through microscopic reversibility

$$\beta_{id} = \alpha_{id} \frac{g_a \Lambda_e^3}{g_i g_e} \exp\left(\frac{I}{k_B T}\right) , \qquad \Lambda_e = h/\sqrt{2\pi m_e k_B T} , \qquad (6.5)$$

where Λ_e denotes the thermal de Broglie wavelength of the electrons and g_e, g_i, and g_a the degeneracy factors of electrons, ions, and atoms, respectively.

Due to interaction effects the rate coefficients become density dependent. The classical screening and quantum effects in dense plasmas decrease the effective ionization energies, $I_{eff} = I - \Delta I$. The ionization coefficients α depend exponentially [EBELING ET AL. 91, EBELING ET AL. 87, EBELING ET AL. 89, KREMP ET AL. 89, BONITZ 90, FÖRSTER 92, SCHLANGES & BORNATH 93] on the lowering of the ionization energy ΔI,

$$\alpha(n_e, n_i, a_a, T) \approx \alpha_{id}(T) \cdot \exp(\Delta I(n_e, n_i, n_a, T)/k_B T) . \qquad (6.6)$$

Therefore interaction effects may change α by orders of magnitudes while recombination coefficients β change only slightly, $\beta \approx \beta_{id}$ [SCHLANGES & BORNATH 93]. This justifies the approximation $\beta = \beta_{id}$. Within the DEBYE approximation the effective ionization energy is given by

$$I_{eff} = I - \frac{\kappa e^2}{4\pi\varepsilon_0} , \qquad \kappa = \sqrt{\frac{2n_e e^2}{\varepsilon_0 k_B T}} , \qquad (6.7)$$

where κ denotes the inverse DEBYE radius. Using the rigid-shift approximation [ZIMMERMANN 88] ΔI can be expressed by the interaction parts of the chemical potentials μ_ν^{int}

$$\Delta I = \mu_a^{int} - \mu_e^{int} - \mu_i^{int} , \qquad (6.8)$$

allowing to include interaction correction beyond the DEBYE approximation. Figure 6.3 shows a schematic representation of the reduction of the ionization energy due to interaction effects.

Diffusion Coefficients for Dense Plasma: The COULOMB interaction prevents the charged particles from diffusing independently [LANDAU & LIFSCHITZ 91b, BONITZ 90]. Therefore local electro-neutrality $n_e(r,t) = n_i(r,t)$ holds in the ambipolar diffusion regime, i.e. for high densities of charge particle. Thus diffusion currents of electrons and ions are the same $j_e(r,t) = j_i(r,t)$. The ambipolar diffusion coefficient D is determined by the elastic ion-atom scattering, $D = 2D_{ia}$, where D_{ia} denotes the binary diffusion coefficient for ions and atoms [BONITZ 90]. An interpolation formula for this quantity was derived by [BONITZ 90]

$$D_{ia} = \frac{3}{8n} \sqrt{\frac{\pi k_B T}{m_a}} J_{ia}(T) , \qquad J_{ia}^{-1} \approx \left(\frac{610}{\sqrt{T/15000K}} - 10 \right) a_B^2 , \qquad (6.9)$$

where n denotes the density of heavy particles, $n = n_i + n_a$. Strictly speaking, non-ideality also affects the diffusion constants [BONITZ 90, OHDE 97] introducing yet another non-linearity in the reaction-diffusion equation. As the focus here is on the comparison of stochastic simulations and analytical results this density effects in the diffusion constant is neglected.

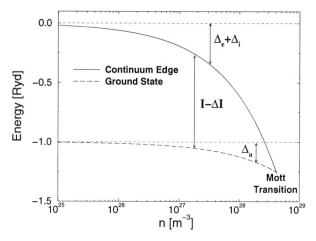

Figure 6.3: Schematic representation of the reduction of the ionization energy due to interaction effects.

Ionization Fronts: The system of coupled partial differential equations (6.1) allows an analytical analysis if the approximation of uniform density of heavy particles is introduced

$$n = n_i(\mathbf{r},t) + n_a(\mathbf{r},t) = \text{constant} . \tag{6.10}$$

Rewriting the reaction-diffusion equations (6.1) in terms of the new variables density of heavy particles $n = n_i + n_a$ and degree of ionization $\bar{z} = n_e/n$ it reduces to [EBELING ET AL. 89]

$$\frac{\partial \bar{z}(\mathbf{r},t)}{\partial t} = w(\bar{z},n,T) + D\Delta\bar{z}(\mathbf{r},t) \tag{6.11}$$

with $w(\bar{z},n,T)$ being the source function. Within the DEBYE approximation one finds

$$w(\bar{z},n,T) = n\alpha_{id}(T)\bar{z}\left[(1-\bar{z})\exp(g(T)\sqrt{\bar{z}n}) - f(T)n\bar{z}^2\right] , \tag{6.12}$$

where

$$g(T) = 2\sqrt{2\pi}\left(\frac{e^2}{4\pi\varepsilon_0 k_B T}\right)^{\frac{3}{2}} , \qquad f(T) = \Lambda^3 \exp\left(\frac{I}{k_B T}\right) . \tag{6.13}$$

While not giving a proper quantitative description of dense hydrogen plasma this model shows several properties found in more sophisticated treatment of the rate coefficients and it still allows to give an analytical expression for the source function.

The general features of equation (6.11) are essentially determined by the roots of the source function $w(\bar{z}, n, T)$, i.e. by its stationary states. One can distinguish two cases (Figure 6.4 [BEULE ET AL. 98b]):

(I) For ideal and for most non-ideal plasmas the source function has two roots, corresponding to one stable stationary state and one unstable stationary state. The stable state $\bar{z}_{s1}$ belongs to the ionization equilibrium at given temperature and density of heavy particles, and the unstable state $\bar{z}_{u1} = 0$ corresponds to an atomic gas with no ionization.

(II) For dense plasmas below a critical temperature, $T < T_c \approx 23000K$ [EBELING ET AL. 89], the source function may have four roots, belonging to two stable and two unstable stationary states. The first unstable state is again the atomic gas $\bar{z}_{u1} = 0$ and the two stable states $\bar{z}_{s1}$ and $\bar{z}_{s2}$ are separated by the second unstable state $\bar{z}_{u2}$ ($\bar{z}_{u1} = 0 < \bar{z}_{s1} < \bar{z}_{u2} < \bar{z}_{s2}$).

Lets assume the system is initially in the unstable state $\bar{z}_{u1} = 0$. This corresponds to the physical situation of an overheated atomic gas. In case (I) the system will undergo a transition from the unstable to the stable state. One scenario is that due to fluctuations ionization spots appear and act as starting points of trigger fronts that propagate in space. In a quasi one-dimensional problem the two stationary state are separated by a moving trigger front which relaxes towards a stable shape $\bar{z}(\chi)$, $\chi = x - v_0 \cdot t$, with velocity, cf. [FÖRSTER 92, MURRAY 93]

$$v_0 = 2\sqrt{D n \alpha_{id}} \ . \tag{6.14}$$

The front shape may be obtained as solution of the boundary-value problem

$$-v\bar{z}' = w(\bar{z}, n, T) + D\bar{z}'' \ , \qquad \bar{z}(-\infty) = \bar{z}_{s1} \ , \qquad \bar{z}(\infty) = \bar{z}_{u1} \ , \tag{6.15}$$

where the prime denotes the derivative with respect to χ. By introducing $y(\bar{z}) = \bar{z}'(\bar{z})$ the boundary-value problem can be transformed into an eigenvalue problem:

$$-vy = w(\bar{z}) + Dyy' \ , \qquad y(\bar{z}_{u1}) = 0 \ , \qquad y(\bar{z}_{s1}) = 0 \ . \tag{6.16}$$

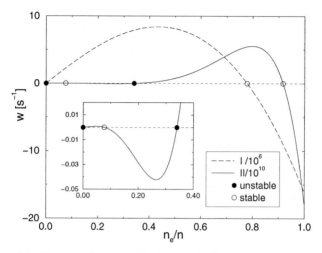

Figure 6.4: The source function $w(\bar{z}, n, T)$ over the degree of ionization $\bar{z}$ for two points in the density-temperature plane: (I) $n = 10^{20}\text{cm}^{-3}$, $T = 35000K$, (II) $n = 0.8 \cdot 10^{22}\text{cm}^{-3}$, $T = 20000K$. The embedded smaller picture shows a magnification of the interval containing the roots $\bar{z}_{u1}$ (left), $\bar{z}_{s1}$ (middle), and $\bar{z}_{u2}$ (right) for case (II). The scaling factors are 10^{-6} and 10^{-10} for case (I) and (II), respectively. Stable stationary states are marked by open circles and unstable ones by filled circles.

The shape of the stable front profile $\bar{z}(\chi)$ is the eigensolution belonging to the smallest eigenvalue v_0 of (6.16) [KOLMOGOROV ET AL. 37, VAN SAARLOOS 89].

In the bistable case (II) there may be two types of fronts. The first one is the same as discussed just before. The second one connects the two stable states $\bar{z}_{s1}$ and $\bar{z}_{s2}$. The velocity and the shape of fronts of this alternative type can again be obtained as solution of the eigenvalue problem (6.16).

So far we have considered only quasi one-dimensional situations. The solution of (6.15) can be generalized for arbitrary dimensions d and the velocity of the front becomes

$$v(K) = v_0 - (d-1)DK , \qquad (6.17)$$

where v_0 denotes again the velocity of plane fronts and K is the local curvature of

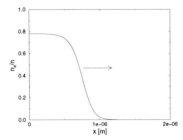 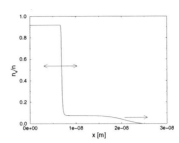

Figure 6.5: Front Propagation: In case (I) the system will undergo a transition from the unstable to the stable state.

Figure 6.6: Front Propagation: In the bistable case (II) there may be two types of fronts.

the front.

6.2   Mesoscopic Description

In order to simulate the reaction-diffusion system (6.1) as a discrete MARKOFF process and thus obtain a mesoscopic decribtion of the inhomogeneous reacting plasma both space and particle number are chosen discrete. To achieve a clear presentation the discussion in this section is limited to the one-dimensional one-component case. First the multi-dimensional state of the mesoscopic description is introduced, then the possible transitions are derived from the elementary physical processes, before summarizing the algorithm that allows to handle the extensive number of possible transitions whose rates may vary by many orders of magnitudes. The details of the algorithm are given in appendix B together with the generalization to higher dimensions and multi-component systems.

Master Equation: Considered a space of length L that is divided into M boxes of length h. The particle density n_i in each box can be represented by a number of particles $N_1, \ldots, N_M$. The refinement of this discretization of particle density can be controlled by an extensive scaling factor Ω,

$$N_i = \Omega n_i . \tag{6.18}$$

The state of the whole system is the direct sum of the local states $\mathcal{N} = \{N_i\}$. The evolution of the number of particles $N_1(t), \ldots, N_M(t)$ is assumed to be a MARKOFF process and thus the dynamics are governed by a master equation for the probability distribution P

$$\dot{P} = \hat{A}P, \qquad P(N_1, \ldots, N_M, t). \tag{6.19}$$

The time-evolution operator $\hat{A}$ consists of a reaction and a diffusion term

$$\hat{A} = \hat{A}_R + \hat{A}_D, \tag{6.20}$$

that can be represented efficiently with the help of the shift operators E_i and E_i^{-1}

$$
\begin{aligned}
E_i F(\ldots, N_i, \ldots) &= F(\ldots, N_i + 1, \ldots) \\
E_i^{-1} F(\ldots, N_i, \ldots) &= \begin{cases} F(\ldots, N_i - 1, \ldots) & \text{if } N_i \geq 0 \\ 0 & \text{if } N_i = 0 \end{cases}
\end{aligned}
\tag{6.21}
$$

that generated or annihilate particle in box i. Within this approch diffusion corresponds to a hopping of particle between neighboring boxes

$$\hat{A}_D = W_D \sum_i^M \left(\hat{E}_{i-1}^{-1} \hat{E}_i + \hat{E}_{i+1}^{-1} \hat{E}_i - 2 \right) N_i. \tag{6.22}$$

while reactions are represented by local birth and death processes, e.g. for the electron equation of (6.1) one yields

$$\hat{A}_R = \sum_i^M W_{ion} \left(\hat{E}_i^{-1} - 1 \right) N_i + W_{rec} \left(\hat{E}_i - 1 \right) N_i(N_i - 1). \tag{6.23}$$

The frequencies W_D, W_{ion}, and W_{rec} for these events have been determined in [BEULE ET AL. 98b] by considering each box as a homogeneous plasma for sufficiently small box lengths h.

Transition Rates: The transition rates for a diffusion diffusion process $R(\mathcal{N} \to \mathcal{N}')$, where

$$\mathcal{N} = \{\ldots, N_i, N_{i+1}, \ldots\} \quad \to \quad \mathcal{N}' = \{\ldots, N_i - 1, N_{i+1} + 1, \ldots\}$$

is approximated by

$$R(\mathcal{N} \to \mathcal{N}') = W_D N_i = \frac{2dD}{h^2} N_i, \tag{6.24}$$

which results in the correct continuum limit.

The considerations will now be specialized for the plasma model with constant density of heavy particles. The source functions $w(\bar{z}, n, T)$ describes an effective transition. Thus one may neglect the chemical origin of the process and derive the effective rates respecting the mean-field equation (6.11). The source function w can be split into a positive part $w^+(\bar{z}, n, T)$ accounting for ionization and a negative part $w^-(\bar{z}, n, T)$ describing recombination:

$$w = w^+ - w^- \quad \text{with} \quad w^\pm \geq 0. \tag{6.25}$$

This two parts will be identified with the birth and death rates of random walkers. In general, the division is not unique, as one can not guarantee, that there is exclusively only one birth or death process. If one adds the same constant both to w^+ and w^-, the mean-field behavior is not affected but the fluctuations increase if the constant is positive and decrease if it is negative.

For the reaction transition rates $R(\mathcal{N} \to \mathcal{N}')$ one obtains

$$\mathcal{N} = \{\dots, N_i, \dots\} \quad \to \quad \mathcal{N}' = \{\dots, N_i + 1, \dots\}$$
$$R(\mathcal{N} \to \mathcal{N}') = W^+ N_i$$
$$\mathcal{N} = \{\dots, N_i, \dots\} \quad \to \quad \mathcal{N}' = \{\dots, N_i + 1, \dots\}$$
$$R(\mathcal{N} \to \mathcal{N}') = W^- N_i$$

If complete ionization $\bar{z} = 1$ is represented by Ω random walkers in a lattice cell, one simply has to multiply (6.11) by Ω and replace the degree of ionization $\bar{z}(r)$ by the quotient

$$\bar{z}(r) = \frac{N_i}{\Omega}. \tag{6.26}$$

Here one takes advantage of the obvious fact that $\bar{z} = n_e/n$ is an intensive concentration. This way one obtains an extensive rate

$$W^+(N_i, n, T) = \Omega w^+\left(\frac{N_i}{\Omega}, n, T\right), \tag{6.27}$$
$$W^-(N_i, n, T) = \Omega w^-\left(\frac{N_i}{\Omega}, n, T\right),$$

from the equation for intensive quantities. Therefore the transition rates are given by

$$R(\mathcal{N} \to \mathcal{N}') = W^+(N_i, n, T), \tag{6.28}$$
$$R(\mathcal{N} \to \mathcal{N}'') = W^-(N_i, n, T),$$

where the states

$$\begin{aligned}
\mathcal{N} &= \{\ldots, N_i, \ldots\}, \\
\mathcal{N}' &= \{\ldots, N_i + 1, \ldots\}, \\
\text{and} \quad \mathcal{N}'' &= \{\ldots, N_i - 1, \ldots\},
\end{aligned}$$

(6.29)

differ by a single birth or decay process.

Markoff Automata: The algorithm used to simulate the master equation is based on a fact that a discrete Markoff process has an exponentially distributed lifetime τ [KARLIN & TAYLOR 75, GILLESPIE 76, GILLESPIE 78, FEISTEL & EBELING 89], see appendix B.3 for details. The probability $P\{\mathcal{N} \to \mathcal{N}'\}$ to enter the new state $\mathcal{N}'$ coming from state $\mathcal{N}$ is proportional to its contribution to leave $\mathcal{N}$

$$P\{\mathcal{N} \to \mathcal{N}'\} = \frac{R(\mathcal{N} \to \mathcal{N}')}{\sum_{\mathcal{N}''} R(\mathcal{N} \to \mathcal{N}'')}.$$

(6.30)

When simulating the master equation for large systems the most time consuming procedure is the probabilistic selection among the large number of possible events. This selection requires a large summation of event probabilities in each simulation step. The problem can be overcome by collecting events into groups: First the probabilities for all events in a certain cell are sumed up. Second cells with probability sums that are similar on a logarithmic scale are collected into classes. Thus the summation of event probabilities can be split into three parts and the selection of the events can be done much more efficiently. After selection and performing a change in particle numbers only a few parts of the sum have to be recalculated in order to proceed with the next selection. The efficiency of the method is based on the fact, that there is only computing time needed, when an elementary process of reaction or diffusion takes place. Therefore, it adapts to the given time scale and is not affected by sharp particle-density gradients which may arise in reacting plasmas. The details of the algorithm are presented in appendix B.

Application: Figure 6.10 shows the simulation of the expansion of a cubic ionization spot into an embedding atomic gas. The degree of ionization in gray scale and contourlines on the faces of a cube. The figure shows how the spot grows and expands into the surrounding atomic gas. The initially sharp boundary between

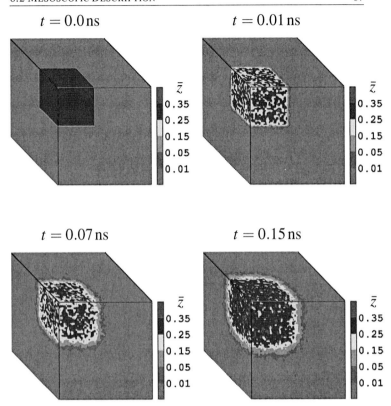

Figure 6.7: Expansion of an ionization spot into an embedding atomic gas. The simulation covers a cube of $50 \times 50 \times 50$ simulation cells (total size $2.5\mu m \times 2.5\mu m.5\mu m$) and the degree of ionization on the faces is represented in gray scale. The initially sharp transition between plasma and atomic gas at $t = 0$ is smeared out, the stable profile develops, and the front propagates into the gas. Both the shape of the spot and the local degree of ionization fluctuate. Extensivety parameter of stochastic simulation $\Omega = 100$, density of heavy particles $n = 10^{20} cm^{-3}$, and temperature $T = 20000K$.

plasma and gas is smeared out and the stable front profile starts to build up. The front propagates while its shape and the degree of ionization fluctuate as indicated by the shades. Due to fluctuations the spot loses its initially symmetry and develops an irregular boundary.

6.3   Ionization Fronts

In order to have an explicit comparison with analytical results we investigate the propagation of a plane ionization front. Consider a plasma with a heavy particles density of $n = 10^{20}$ cm^{-3} and a temperature of $T = 35000$ K. This plasma is non-degenerate, strongly coupled, and an example of case (I), i.e., with two stationary states. The stable state corresponds to an ionization of $\bar{z}_{s1} = 0.779$ while the unstable one belongs to the atomic gas, $\bar{z}_{u1} = 0$. The length of the simulation cells is chosen $h = 2.5 \cdot 10^{-8}$ m to be much larger than the screening length κ^{-1}, justifying the assumption of electro-neutral cells and ambipolar diffusion. One single cell of this size contains approximately 1600 atoms or ions. We start out from a rectangular shaped region with ionization $\bar{z}_{s1}$ embedded in a surrounding atomic gas, see Figure 6.8 [BEULE ET AL. 98b]. The simulation shown in this figure covers a total of 2000 cells using periodic boundary conditions and an extensivety parameter $\Omega = 100$. As time evolves the initially ionized region ($t = 0$ ns in Figure 6.8) starts to expand into the gas. After some time ($t = 0.12$ ns in Figure 6.8) the stable front shape has almost developed. The front propagates with constant shape and velocity into the atomic gas ($t = 2.77$ ns in Figure 6.8), until the whole area is in the state of equilibrium ionization ($t = 4.20$ ns in Figure 6.8). At any time one can see the fluctuations of the degree of ionization.

Measuring the Front Velocity: In the macroscopic description a stable front solution[1] $\tilde{c}$ is defined by the fact that it is stationary $\tilde{c}(u,t) = \tilde{c}(u)$ in the reference frame $u = x - vt$ that moves with the front velocity v. Obviously this definition can not be utilized in the stochastic description. In order to investigate the propagation of unstable or fluctuating front shapes a pragmatical measure for the front position

---

[1]For simplicity we introduce the normalized degree of ionization $c(x,t) = z(x,t)/z_{s1}$ which becomes one in the stable state.

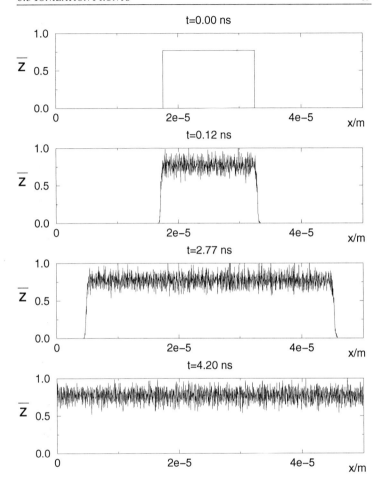

Figure 6.8: Development of two plane fronts starting out from resting rectangular shape (upmost picture). The boundaries begin to move to the left and to the right into the area of atomic gas (two central pictures) until the ionization equilibrium is reached everywhere (lowest picture). The degree of ionization is drawn over the spatial coordinate. Density of heavy particles $n = 10^{20}$ cm^{-3}, temperature $T = 35000$ K.

at time t can be used

$$c_f(t) = \int\limits_{x_0}^{+\infty} c(x,t)\,\mathrm{d}x\,, \qquad (6.31)$$

where x_0 is a point far enough left of the initial front position and $c_f(t)$ gives the average value of the quantity c in the half space $x > x_0$ at time t. The velocity of the propagating front is now characterized by

$$v_f(t) = \frac{\mathrm{d}c_f(t)}{\mathrm{d}t} = \frac{\mathrm{d}}{\mathrm{d}t} \int\limits_{x_0}^{+\infty} \left(\gamma \frac{\partial^2 c}{\partial x^2} + w(c) \right) \mathrm{d}x = \int\limits_{x_0}^{+\infty} w(c)\,\mathrm{d}x. \qquad (6.32)$$

For stable front solution this becomes equivalent to the former definition, cf. [MURRAY 93, BREUER ET AL. 94]. For the mesoscopic picture the definition of the front position corresponding to (6.31) is given by

$$N_{tot}(t) = \sum_{i=1}^{M} N_i(t)\,. \qquad (6.33)$$

The time derivative of this quantity

$$v_m = \frac{h}{z_{s1}\Omega} \frac{\mathrm{d}\langle N_{tot}\rangle}{\mathrm{d}t}\,, \qquad (6.34)$$

provides a definition of wave front velocity of the stochastic process without invoking the concept of stationarity in a co-moving frame.

Plane Ionisation Fronts: As mentioned before, we have to ensure that the extensivety parameter Ω is chosen large enough to give proper results. We check this by comparing the stable velocities obtained from simulations with different Ω with the analytically result from equation (6.14), see Table 6.1 [BEULE ET AL. 98b]. The velocities were obtained by averaging over a long period of time which was chosen such that the statistical error of the velocities is about 50 m/s. It can be seen that already a small extensivety parameter $\Omega = 50$ is sufficient to obtain a reasonable approximation while $\Omega \geq 200$ leads to an excellent agreement with the theoretical prediction. For a more thorough discussion of the dependency of the front speed on the extensivety parameter and fluctuation effects cf. section 6.4 and [MAI ET AL. 98, BRUNET & DERRIDA 01, WARREN ET AL. 01].

Ω	50	100	200	500	theory
v_0	4460 m/s	4540 m/s	4760 m/s	4760 m/s	4818 m/s

Table 6.1: Dependency of the front velocity on the extensivety parameter Ω

Front shape: In Figure 6.8 all 2000 cells of the simulation at a time. A more detailed picture covering only 80 cells is given in Figure 6.9 [BEULE ET AL. 98b], where the numerical solution of the eigenvalue problem (6.16) is drawn for comparison. These two results for the front shape are also in very good agreement.

Simulations were also performed in the bistable region of the model plasma at $T = 20000$ K and $n = 0.8 \cdot 10^{22}$ cm^{-3}. The front type which is present in both cases (I) and (II), as well as the alternative front type that connects the two stable states were observed.   Again a stable front profile develops and shape as well as velocity are in good agreement with the solution of the eigenvalue problem (6.16). However the shape is unrealistically sharp and the velocity is unrealistically high in comparison to inter-particle distance and thermal velocity, respectively. As several approximations had to be introduced in order to gain an analytically tractable model, we can not expect to give a proper physical description in the whole density-temperature plane. In the bistable region the Debye approximation leads to an overestimation of non-ideality.

Curved Ionization Fronts   For the simulation of a curved ionization front the density of heavy particles and the temperature are chosen as in the first example. The simulation is performed on a grid of 200×200 cells giving a total area of 5μm $\times 5\mu$m. The simulation uses $\Omega = 100$ random walkers per cell. Initially a circular spot with a diameter of 1.5μm is set to the equilibrium ionization state.

Figure 6.10 [BEULE ET AL. 98b] shows how the spot grows and expands into the surrounding atomic gas. The initially sharp boundary between plasma and gas is smeared out and the stable front profile starts to build up. The front propagates while its shape and the degree of ionization fluctuate as indicated by the shades that are most clearly seen on the lower pictures of Figure 6.10. Due to fluctuations the spot loses its initially spherical symmetry and develops an irregular boundary with different local curvature.

In Figure 6.11 [BEULE ET AL. 98b] we compare the mean velocity of the spot

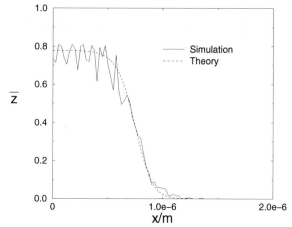

Figure 6.9: Snapshot of the developed front shape in the stochastic simulation (solid line) with $\Omega = 200$ together with the solution of the eigenvalue problem (dashed line). The degree of ionization is drawn over the spatial coordinate. Density of heavy particles $n = 10^{20}$ cm^{-3}, temperature $T = 35000$K.

front in the simulation with the theoretical value given by (6.17), where we use $v_0 = 4540$m/s as obtained for the same value of $\Omega = 100$ (Tab. 6.1). The initially rectangular front has a curvature of $K = 1.33 \mu$m^{-1}. While the front profile approaches its stable shape the front velocity increases from $v = 0$ towards the curvature-depending value predicted by (6.17). This value is reached for a spot radius of approximately 1.8μm or after a time of about 0.3 ns. Afterwards the mean front velocity fluctuates about the predicted value and increases slowly as the curvature decreases with the expansion of the spot.

As we have found very good agreement in the determination of key quantities it is justified to proceed to more realistic plasma models [EBELING ET AL. 76, KRAEFT ET AL. 86, FORTOV & IAKUBOV 90, EBELING ET AL. 91, EBELING ET AL. 92, ICHIMARU 94, KOBZEV ET AL. 95, KRAEFT & SCHLANGES 96] which go beyond the Debye screening and account for quantum effects as well as many-particle interaction. As the method of

$t = 0.007\,\text{ns}$ $t = 0.103\,\text{ns}$

$t = 0.250\,\text{ns}$ $t = 0.320\,\text{ns}$

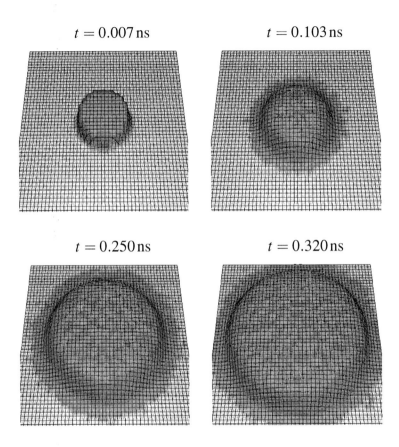

Figure 6.10: Expansion of an ionization spot into an embedding atomic gas. The distance of grid lines is 10^{-7} m, the whole picture shows an area of $5\,\mu\text{m} \times 5\,\mu\text{m}$. The height represents the degree of ionization on a linear scale. The initially sharp transition between plasma and atomic gas at $t = 0$ is smeared out, the stable profile develops, and the front propagates into the gas. Both the shape of the spot and the local degree of ionization fluctuate. Extensivety parameter of stochastic simulation $\Omega = 100$, density of heavy particles $n = 10^{20}$ cm^{-3}, and temperature $T = 35000$ K.

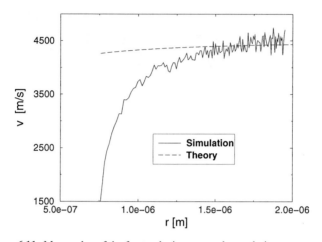

Figure 6.11: Mean value of the front velocity averaged at each time moment over the whole spot boundary (solid line) and curvature-dependent theoretical values (dashed line) plotted over mean spot radius. Extensivety parameter of stochastic simulation $\Omega = 100$, density of heavy particles $n = 10^{20} \mathrm{cm}^{-3}$, and temperature $T = 35000K$.

MARKOFF automata is capable of handling multi-component systems it will be possible to give up the approximation of constant density of heavy particles and study both the simultaneous fluctuations of total density and degree of ionization. We expect that this will allow for a kinetic description of the plasma phase transition and the phase separation in the coexistence area of the PPT. Since there are no principle restrictions with respect to the dimensionality and geometry of the problem also various experimental setups can be treated with our method.

Conclusion: Ionization kinetics and front propagation in partially ionized dense plasmas is governed by reaction-diffusion equations. We have shown the method of Markoff automata to be an excellent tool to investigate this kinetic properties of inhomogeneous reacting dense plasma by comparing stochastic simulations and analytical results.

6.4   Fluctuation Effects on Front Propagation

It was already pointed out that each stochastic simulation develops a front shape that compares well with the analytical prediction (Figure 6.9) and that the mean front velocity is also in good agreement with the analytical results (see Table 6.1). Nevertheless the front position of different stochastic realization that evolved from the same initial conditions differ significantly from each other and from the results for the mean-field equation, see Figure 6.12.

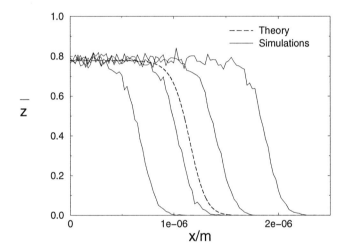

Figure 6.12: Snapshot of different stochastic realization that evolved from the same initial conditions (rectangular front) with analytical results for the fronts shape of the mean-field equation (6.11) at arbitrary position. Density of heavy particles $n = 10^{20}\,\mathrm{cm}^{-3}$, temperature $T = 35000\,K$, extensivity parameter $\Omega = 200$.

In order to understand this effect one has to analyse how the mean-field equation (6.11) is related to the mesoscopic describtion (6.19). Usually one argues that fluctuations are small and therefore it suffices to investigate the dynamics of the average. However this is true only if fluctuations are damped, i.e. if the macroscopic dynamics is asymptotically stable. Under this condition the system size expansion provides a unique decomposition of the stochastic process into a macroscopic part which is governed by a deterministic differential equation and small gaussian fluctuations around the macroscopic dynamics [VAN KAMPEN 92]. It turns out that this prerequisite for the system size expansion is violated for equation (6.11) and thus a non-perturbative fluctuation effect on the profile position is observed. This effect does not depend crucially on the interaction effects but rather affects all wave fronts that propagate into an unstable stationary state. Therefore it is reasonable to consider the analytical results on front stability for the ideal limit, while the numerical determination of the corresponding scaling law is based on the full model including interaction effects.

Results on Front Stability: For a non-interacting partially ionized model plasma the reaction-diffusion equation (6.11) is reduced to the well known FISHER-KOLMOGOROV type with a quadratic source term. Using appropriate scaling this equation can be rewritten as

$$\frac{\partial c}{\partial t} = \gamma \frac{\partial^2 c}{\partial x^2} + w(c) = \gamma \frac{\partial^2 c}{\partial x^2} + c - c^2 \tag{6.35}$$

and admits wave-front solutions propagating between a stable stationary $c \equiv 1$ and an unstable stationary state $c \equiv 0$. Macroscopic studies [VAN SAARLOOS 89] have lead to the conclusion that there exist a continuum of possible propagation velocities greater than the minumum velocity $v_{min} = 2\sqrt{\gamma}$. The analytical form of the (marginally) stable wave solution could not be determined analytically so far but the front width can be characterized by the front steepness at the inflection points. Namely $\sqrt{\gamma}$ is proportional to the front width for the front moving with v_{min}. It has been shown that sufficiently steep initial profiles evolve to the front propagating at the minimum allowed velocity [VAN SAARLOOS 89].

Many studies have investigated the derivations from the the mean-field picture due to fluctuation effects present at the microscopic level [BREUER ET AL. 94, RIORDAN ET AL. 95, MAI ET AL. 96, KESSLER ET AL. 98, MAI ET AL. 00, WARREN ET AL. 01, BRUNET & DERRIDA 01]. It has become clear (cf. [BRUNET & DERRIDA 01]) that in mesoscopic and microscopic approaches a single velocity is selected for arbitrary initial conditions, the velocity is usually smaller than v_{min} and tends slowly towards the mean-field limit as Ω increase. The front shape fluctuates but does not broden with time, and the front position diffuses with time, with a diffusion constant which decrease slowly with Ω, Scaling laws for front propagation speed and diffusion constants have been proposed in [WARREN ET AL. 01, BRUNET & DERRIDA 01]. Fluctuations effects are not restricted to the one dimensional case considered here but also influence front propagation in higher dimensional systems [RIORDAN ET AL. 95, MAI ET AL. 98]. In the limit of small concentrations the continuum modeling breaks down [MAI ET AL. 96] and the front propagation is governed mainly by fluctuations.

System size expansion: The proper choice of the simulation cells size h has to balance between two opposite constraints. On the one hand, the cell must be sufficiently small to be considered homogeneous, on the other hand, a proper description of the chemical reactions by birth and death processes requires sufficiently large cells containing a large number N of particles. The later condition also justifies a systems size expansion of the master equation to the first order of $1/N$ [VAN KAMPEN 92]. The system size expansion[2] establishes the relation between mesoscopic and macroscopic description by assuming that the number of particles in each box N_i can be represented as $N_i = \Omega c_i + \sqrt{\Omega}\, \eta_i$ where η_i denote the fluctuation about the mean values c_i. By using this expansion the master equation for the distribution of particle numbers $P(N_1, \ldots, N_M, t)$ is transformed into a differential equation for the densities and a FOKKER-PLANCK equation for the distribution of the fluctuations $\Pi(\eta_1, \ldots, \eta_M, t)$. The system size expansion requires that fluctuations are always damped in order to keep the distribution of the fluctuations unimodal and narrow. In the model (6.19) this requirement is violated by fluctuations that correspond to a movement of the whole front relative to a frame moving with the average front velocity $\bar{v}$. Thus the prerequisites for the system size expansion is violated and non-perturbative fluctuation effects that obey no simple macroscopic law can occur. The undamped GOLDSTONE modes reflect the translation invariance of the system [VAN KAMPEN 92] and lead to a diffusion-type behavior of the front position [BREUER ET AL. 94]. Therefore the position of different stochastic realization of ionization fronts starting from the same initial conditions will drift apart, i.e. the distribution of front position is still unimodal and gaussian but no longer narrow.

Scaling: Fluctuations are usually predicted to be of order $\Omega^{0.5}$ for the stable case or of order Ω^1 for diffusion-type behavior. In order to quantify the fluctuations of the front position one has to consider the variance of the front position (6.31) and

---

[2]Also named Ω-expansion.

(6.33) defined by

$$Var(N_{tot})(t) = \frac{1}{k-1} \sum_{j=1}^{k} (N_{tot}^{(j)}(t) - \bar{N}_{tot}(t))^2 . \qquad (6.36)$$

Here index j labels k different mesoscopic realisation. After a certain relaxation time that corresponds to the relaxation of the front shape and average velocity the variance $Var(N_{tot})$ grows linear with time [BREUER ET AL. 95]

$$\hat{D} = \frac{\Delta Var(N_{tot})}{\Delta t} \qquad (6.37)$$

as expected for a diffusive process with diffusion constant $\hat{D}$. The relative spread of the front position is given by

$$\frac{\sqrt{Var(N_{tot})}}{\langle N_{tot} \rangle} \approx \frac{\sqrt{\hat{D}t}}{\bar{v}_m t} \qquad (6.38)$$

where $\langle N_{tot} \rangle = \bar{v}_m t$ was used. Using a power law $\hat{D} = b\Omega^{2a}$ for the scaling behavior and $\bar{v}_m t \approx \Omega t$ this result can be rewritten as

$$\frac{\sqrt{Var(N_{tot})}}{\langle N_{tot} \rangle} \approx \frac{\sqrt{\hat{D}t}}{\bar{v}_m t} \sim \frac{\sqrt{\Omega^{2a}}}{\sqrt{t}\,\Omega} = \frac{1}{\sqrt{t}}\Omega^{a-1} , \qquad (6.39)$$

The front-position diffusion has been estimated from ten realisations of system (6.11) for the following extensivity parameters $\Omega = 30, 100.300, 1000, 3000, 10000, 30000$. For each Ω the temporal development of $\sqrt{Var(N_{tot}(t))}$ was fitted to $\sqrt{\hat{D}t}$ in order to estimate the front-diffusion coefficient $\hat{D}$. The scaling law can be determined by a least square fit of $\hat{D}(\Omega) = b\Omega^{2a}$. This procedure yields $a = 0.75 \pm 0.12$ which differs significantly from normally damped fluctuations $a = 0.5$ as well as from an undamped behavior $a = 1$. For the FISHER-KOLMOGOROV with a simple quadratic source function, which corresponds to the ideal plasma limit [BREUER ET AL. 95] found a comparable value $a = 0.84 \pm 0.02$.

Conclusions: The effects of fluctuation on front propagation can be summarized as follows: fluctuations in direction of front propagation show an unusual scaling behavior and are only weakly damped. Fluctuations in perpendicular to the direction of front propagation are damped normally $\sim \Omega^{-0.5}$ (front shape). Since the resulting exponent $a - 1$ is smaller than zero the relative spread of the front position vanishes in the macroscopic limit. But the dynamics of the front approaches the deterministic limit extremly slowly. Due to this character it is advantageous to measure the front velocity from an ensemble average rather than from a time average.

6.5 Multi-Component Systems

The carbon-hydrogen plasma studied in section 3.2 forms a complex reacting multi-component system. Here we study the recombination of an initially completely ionized homogeneous carbon-hydrogen plasma. The densities of carbon and hydrogen are chosen 10^{22}cm^{-3} and $2 \cdot 10^{22}\text{cm}^{-3}$, respectively. Due to the temperature of 100 000K hydrogen stays almost completely ionized at any time. Atomic states of carbon with the same electron configuration have very similar energies [MOORE 93] and are collected into one level. The thermodynamic equilibrium state consists mainly of C^{4+} ions. Therefore one has 8 kinds of reacting constituents (electrons, bare nuclei C^{6+}, and ions C^{5+}, C^{4+} with ground state and two excited levels each) with 36 reactions which are considered using semiempirical cross sections, cf. [BEULE ET AL. 95]. Figure 6.13 [BEULE ET AL. 96] shows a comparison of stochastic and deterministic kinetics. The rate equation of deterministic kinetics were integrated with a forth-order Runge-Kutta procedure. The master equation of stochastic kinetics have been simulated for an ensemble of $N_C = 1000$ carbon and 2000 Hydrogen ions. The stochastic realisations fluctuate around the deterministic curves. For further discussion of this system cf. [BEULE ET AL. 95].

6.6 Temperature Equations

So far we have considered only isothermal processes. For real plasmas this will not always be appropriate. Therefore the kinetics are now combined with a temperature equations that allows to fulfill various thermodynamic constraints. If volume and total number of heavy constituents (i.e. density) in the plasma are kept constant, isothermal relaxation, $dT/dt = 0$, leads to a minimum of the free energy F with respect to the composition $\{N_v\}$. In the deterministic description the change

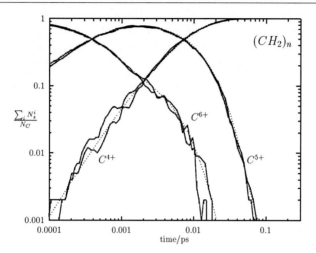

Figure 6.13: The time evolution of relative abundances of carbon ions C^{6+}, C^{5+}, and C^{4+} obtained from deterministic (dotted lines) and stochastic kinetics (solid lines, two realisations).

in, e.g. entropy during relaxation can be calculated from

$$\frac{dS}{dt} = -\sum_\nu \frac{\partial \mu_\nu}{\partial T}\bigg|_{V,\{N_{\gamma \neq \nu}\}} \frac{dN_\nu}{dt} . \tag{6.40}$$

where the change in composition dN_ν/dt is given by the rate equations. For the stochastic approach an equivalent discrete procedure has to be used

$$\frac{\Delta S}{\Delta t} = -\sum_\nu \frac{\Delta \mu_\nu}{\Delta T}\bigg|_{V,\{N_{\gamma \neq \nu}\}} \frac{\Delta N_\nu}{\Delta t} , \tag{6.41}$$

where the change in composition $\Delta N_\nu/\Delta t$ is now given by the master equation [BEULE ET AL. 96].

Under isentropic conditions, $dS(\{N_\nu\}, T, V)/dt = 0$, the internal energy U is minimized with respect to the composition N_ν. The changes in composition leads to a change in energy and temperature given by

$$\frac{dU}{dt} = \sum_\nu \left[\frac{\partial U}{\partial N_\nu}\bigg|_{V,T,\{N_{\gamma \neq \nu}\}} - T\frac{\partial \mu_\nu}{\partial T}\bigg|_{V,\{N_\gamma\}} \right] \frac{dN_\nu}{dt} , \tag{6.42}$$

$$\frac{dT}{dt} = -\frac{\partial T}{\partial S}\bigg|_{V,\{N_\gamma\}} \sum_\nu \frac{\partial \mu_\nu}{\partial T}\bigg|_{V,\{N_\gamma\}} \frac{dN_\nu}{dt} .$$

Relaxation under isoenergetic conditions, $dU(\{N_v\}, T, V)/dt = 0$, corresponds to maximization of the entropy with respect to composition. Changes in entropy and temperature can be derived in analogy to equations 6.40 and 6.42. The contineous version of the temperature equation (6.42) has been derived directly from quantum statistics in DEBYE approximation [OHDE ET AL. 95, OHDE ET AL. 96]. Here the interaction effects in the rate equation are treated using equation given in section 6.2 and the PADÉ approximation given in section 2.5.

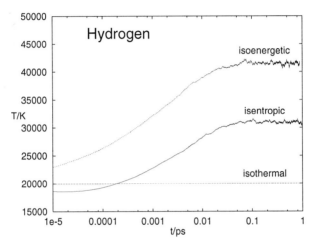

Figure 6.14: Time evolution of the temperature in hydrogen plasma under isentropic, isothermal, and isoenergetic conditions.

Figures 6.14 to 6.16 [BEULE ET AL. 96] compare the isothermal, isoenergetic, and isotropic recombination of an initially completely ionized hydrogen plasma. For the chosen density of $n = 5 \cdot 10^{21} \mathrm{cm}^{-3}$ and the initial temperature $T = 30\,000\mathrm{K}$ only one bound state is present. Isoenergetic as well as isentropic conditions lead to a higher final temperature then isothermal recombination (Fig. 6.14) and, therefore, to a higher degree ionization (Fig. 6.15). To achieve isothermal conditions large amounts of entropy have to be exported from the plasma on very short time scales (Fig. 6.16). In each of the considered situations, non-ideality effects lead to an increase in relaxation time and a higher degree of ionization that corresponds to pressure ionization in equilibrium state.

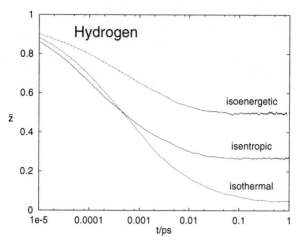

Figure 6.15: Time evolution of the degree of ionization in hydrogen plasma under isentropic, isothermal, and isoenergetic conditions.

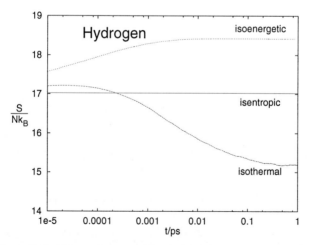

Figure 6.16: Time evolution of the reduced entropy in hydrogen plasma under isentropic, isothermal, and isoenergetic conditions.

Summary

The equation of state of hydrogen and deuterium at high pressure is still uncertain in several important areas. Shock wave experiments by [COLLINS ET AL. 98] and [KNUDSON ET AL. 01] show conflicting results at about 100 GPa and revealed some significant discrepancies to standard equation of state libraries. Furthermore the physical nature of the insulator-to-metal transition in hydrogen, that has been verified experimentally for the first time by [WEIR ET AL. 96] is still controversial.

In this work the thermodynamic properties of dense plasma are investigated using the technique of PADÉ approximations of within the chemical picture (PACH). The expressions given by [EBELING 90, STOLZMANN & BLÖCKER 96] are improved with new interpolation formulas for the quantum virial functions and used to determine the adiabatic equation of state of a partially ionized hydrogen plasma and the detailed composition of the carbon-hydrogen plasma in a wide density and temperature range. Utilizing improved data for the dense neutral fluid a new equation of state for dense hydrogen plasma is constructed which gives a realistic behavior for low (neutral molecular fluid) and high temperatures (highly ionized plasma). Compared with previous approaches the correlations in the dense, neutral component are considered in a systematic way. A thermodynamic instability occurs in the intermediate region that is connected with a first-order phase transition. This so far hypothetical *plasma phase transition* describes a transition from a partially dissociated fluid to a partially ionized plasma and is closely related to the insulator-to-metal transition. The corresponding coexistence line and the coexistence region were derived for temperatures from 2000 to 10000 K. In this range the coexistence pressure varies between 120 and 110 GPa and the phases differ by about $0.2 \, \text{g/cm}^3$ in mass density. The electrical conductivity determined on the basis of the detailed composition of the PACH model qualitatively reproduces the experimentally observed behavior, i.e. a transition from a

semi-conducting behavior to metal-like conductivity. The HUGONIOT curves of
deuterium and hydrogen have been determined with initial condition corresponding to shock wave experiments. Furthermore the influence of interaction effects
and bound state formation on the asymptotic behavior of the compression could
be determined. For high temperatures the HUGONIOT curves obtained within the
PACH model is in good agreement with the analytical results as well as with the
activity expansion [ROGERS & YOUNG 97], and path integral Monte-Carlo simulations [MILITZER & CEPERLY 00]. Using the improved treatment of the neutral
component in the low temperature region one finds good agreement with the experimental data of [COLLINS ET AL. 98] as well as with the linear mixing model of
[ROSS 98]. The calculations presented here within the PACH approach show that
the large increase in density between 30 and 100 GPa observed in the single shock
experiments as well as the conductivity measurements are consistent with the existence of a plasma phase transition. Given the peculiar experimental situation that
is reflected in the large uncertainties in the current experimental data it must be
concluded that neither the single shock nor the multiple shock experiments allow
to finally exclude or confirm the existence of a first-order plasma phase transition.
Due to the proximity of single shock HUGONIOT curve and the predicted critical
point multiple shock experiments seem more appropriate to clarify the situation.

Besides equilibrium properties this work also deals with the kinetics of electron transitions in partially ionized dense plasmas, which can be approximated by
reaction-diffusion equations. A stochastic replacement for the deterministic differential equations based on a master equation is proposed. In this mesoscopic
approach the particles are presented by random walkers that perform hopping processes on a lattice resulting in an effective diffusion and birth and death processes
representing ionization and recombination. The stochastic description forms the
basis of an efficient simulation technique which is applied to a multi component
system, two- an three-dimensional geometries and in combination with a temperature equation. A comparison with analytic results for profile and velocity of ionization fronts confirms the stochastic simulation method. Fluctuation effects generate
a diffusive instability of the front position. This fluctuation effect is only weakly
damped during the transition from the mesoscopic to the deterministic description and thus confirms the scaling behavior found in simpler reaction-diffusion
systems.

Zusammenfassung

Die Zustandsgleichung von Wasserstoff bzw. Deuterium bei großem Druck ist in wichtigen Bereichen immer noch unklar. Stoßwellenexperimente von [COLLINS ET AL.98] und [KNUDSON ET AL.01] ergaben im Bereich von etwa 100 GPa widersprüchliche Ergebnisse und teilweise große Abweichungen zu den Standardbibliotheken für Zustandsgleichungen. Darüber hinaus ist die physikalische Natur des von [WEIR ET AL.96] erstmals experimentell nachgewiesenen Isolator-Metallübergangs immer noch strittig.

In dieser Arbeit werden die thermodynamischen Eigenschaften dichter Plasmen mit der Methode der PADÉ-Näherungen im chemischen Bild (PACH) untersucht. Die von [EBELING 90, STOLZMANN & BLÖCKER 96] vorgeschlagenen Ausdrücke werden mit einer neuen Interpolationsformel für die Quantenvirialfunktionen verbessert und zur Bestimmung der adiabatischen Zustandsgleichung für partiell ionisiertes Wasserstoffplasma, sowie die detaillierte Zusammensetzung von Kohlenwasserstoffplasma für einen weiten Dichte-Temperaturbereich angewendet. Mit verbesserten Daten für das dichte neutrale Fluid wird ein Zustandgleichungsmodell für dichten Wasserstoff erstellt, welches ein realistisches Verhalten beschreibt: Für niederige Temperaturen ein neutrales molekuares Fluid und für hohe Temperaturen ein hoch ionisiertes Plasma. Verglichen mit früheren Ansätzen werden die Korrelationen im dichten neutralen Fluid systematisch berücksichtigt. Das Modell zeigt eine thermodynamische Instabilität, die mit einem Phasenübergang erster Ordnung verknüpft ist. Dieser noch nicht verifizierte *Plasmaphasenübergang* beschreibt den Übergang von einem teilweise dissoziierten Fluid zu einem teilweise ionisierten Plasma und ist eng verkünpft mit dem Isolator-Metallübergang. Die Koexistenzlinie und der Koexistenzbereich wurden für Temperaturen von 2 000 bis 10 000 K bestimmt. In diesem Bereich liegt der Koexistenzdruck zwischen 120 und 110 GPa und die Dichte der Phasen unterscheidet sich um ca. $0.2\,\mathrm{g/cm^3}$. Die auf Basis der detaillierten Zusammensetzung im PACH-Modell berechnete elektrische Leitfähigkeit reproduziert qualitativ das experimentell gemessene Verhalten, d.h. einen Übergang von einer halbleiterartigen zu einer metallähnlichen Leitfähigkeit. Die HUGONIOT-Kurven von Wasserstoff bzw. Deuterium wurden für die in den Stoßwellenexperimenten verwendeten Anfangsbedingungen untersucht. Außerdem konnte eine analytische Beschreibung der Einflüsse von Wechselwirkungseffekten und der Bildung gebundener Zustände auf das asymptotische Verhalten der Kompression gefunden werden. Die mit der PACH-Methode gewonnenen HUGONIOT-Kurven zeigen im Hochtemperaturbereich eine gute Übereinstimmung mit den analytsichen Ergebnissen, der *activity expansion* [ROGERS & YOUNG 97] und Pfadintegral-Monte-Carlo Simulationen [MILITZER & CEPERLY 00]. Mit der verbesserten Behandlung der ungeladenen Komponenten, findet man im Niedertemperaturbereich eine Übereinstimmung mit experimentellen Daten aus den Stoßwellenexperimenten von [COLLINS ET AL. 98] sowie mit dem *linear mixing model* [ROSS 98]. Diese Berechnungen mit dem PACH-Ansatz zeigen, dass der große Dichteanstieg, der in den Einzelstoßexperimenten zwischen 30 and 100 GPa beobachtet wird, ebenso wie die Leitfähigkeitsmessungen konsistent mit der Existenz eines Plasmaphasenübergangs sind. Aufgrund der schwierigen experimentellen Situation, die sich in den großen Unsicherheiten der experimentellen Ergebnisse niederschlägt, läßt sich die Existenz eines Plasmaphasenübergangs erster Ordnung immer noch nicht sicher bestätigen oder ausschließen. Wegen der Nähe der HUGONIOT-Kurven der Einzelstoßexperimente zum vorhergesagten kritischen Punkt scheinen Multistoßexperimente besser geeignet, die Situation entgültig aufzuklären.

Neben den Gleichgewichtseigenschaften befaßt sich diese Arbeit auch mit der Kinetik von Elektronenübergängen in partiell ionisierten dichten Plasmen, welche näherungsweise durch Reaktions-Diffusiongleichungen beschrieben werden können. Es wird eine stochastische Beschreibung als Alternative zur deterministischen Beschreibung mit Differentialgleichungen vorgeschlagen. Bei diesem mesoskopischen Herangehen unterliegen die Teilchen einer Zufallsbewegung in einem diskreten Raum, welche zu einer effektiven Diffusion führt. Ionisation und Rekombination werden durch Entstehungs- und Vernichtungsprozesse beschrieben. Die stochastische Beschreibung ist die Grundlage für ein effektives Simulationsverfahren, welches für Vielkomponentensysteme, zwei- und dreidimensionale Systeme und in Kombination mit einer Temperaturgleichung angewendet wird. Durch

Vergleich mit analytischen Ergebnissen für die Profile und Geschwindigkeiten von Ionisationsfronten wird die stochastische Methode validiert. Die Frontposition ist in der mesoskopischen Beschreibung instabil und unterliegt einem Diffusionprozess. Beim Übergang von der mesoskopischen zur deterministischen Beschreibung wird dieser Fluktuationseffekt nur schwach gedämpft und bestätigt damit das in einfacheren Reaktions-Diffusionssystemen gefundene Skalierungsverhalten.

A Limiting Laws and Interpolation Formulas

A.1 Fermi Integrals

In the treatment of fermion systems at finite temperatures, it is useful to define the FERMI integrals e.g. [ZIMMERMANN 88],

$$I_j(x) = \frac{1}{\Gamma(j+1)} \int_0^\infty \frac{t^j}{\exp(t-x)+1} \, \mathrm{d}t \qquad \text{for} \quad j > -1 \qquad (A.1)$$

where the prefactor containing the GAMMA function ensures simple asymptotic behavior $I_j(x) \to exp(x)$ in the classical limit $x \to -\infty$. For the highly degenerate case one finds

$$I_j(x) \approx \frac{x^{j+1}}{\Gamma(j+2)} \qquad \text{if} \quad x \gg 1 \,. \qquad (A.2)$$

For 3-dimensional fermion systems the dimensionless argument $x = \mu/(k_B T)$ of equation (A.1) is defined through the relation

$$I_{\frac{1}{2}}\left(\frac{\mu}{k_B T}\right) = \frac{n\Lambda^3}{g} \,, \qquad (A.3)$$

where g denotes the degeneracy and the free energy is given by

$$f = -k_B T \frac{g}{n\Lambda^3} I_{\frac{3}{2}}\left(\frac{\mu}{k_B T}\right) + n\mu \,. \qquad (A.4)$$

Other thermodynamic functions can be constructed by means of the differentiation rule

$$\frac{\mathrm{d}I_j(x)}{\mathrm{d}x} = I_{j-1}(x) \,. \qquad (A.5)$$

In order to calculate any thermodynamic functions for a given density and temperature e.g. $f(n,T)$ one needs to invert equation (A.3) first. However, in practical calculations one prefers analytical expressions in terms of density and temperature.

Interpolation Formula: A parametrization of the inverse function of $I_{1/2}$ can be found in [ZIMMERMANN 88],

$$I_{1/2}^{-1}(y) = \begin{cases} \ln y + 0.3536\, y - 0.00495\, y^2 + 0.000125\, y^3 & \text{if } y < 5.5 \\ 1.209\, y^{\frac{2}{3}} - 0.6803\, y^{-\frac{2}{3}} - 0.85\, y^{-2} & \text{if } y \geq 5.5 \end{cases} \tag{A.6}$$

where it is used to establish very accurate analytical approximation of the integrals $I_{1/2}$ and $I_{3/2}$ as functions of $n\Lambda^3/g$, e.g. $I_{3/2}(y)$

$$I_{3/2}(y) = \begin{cases} \frac{y}{1+0.3536y-0.0099y^2+0.000375y^3} & \text{if } y < 5.5 \\ \frac{y^{1/3}}{0.806+0.4533y^{-4/3}+1.7y^{-8/3}} & \text{if } y \geq 5.5 \end{cases} \tag{A.7}$$

A.2 One Component Plasma

The classical one component plasma (OCP) is a prototype dense-plasma model, where the ions are point charges while the electrons form a rigid uniform background. The OCP in its fluid and crystal state has been subject of extensive Monte-Carlo [STRINGFELLOW ET AL. 90, DEWITT 76], molecular dynamic simulations [BAUS & HANSEN 80] and path integral Monte-Carlo simulations [JONES & CEPERLEY 96]. These allow to assess the validity of limiting laws for weak coupling [DEBYE & HÜCKEL 23, ABE 59, COHEN & MURPHY 69] and more approximative integral equations theories [ICHIMARU 94] like hyper-netted chain (HNC). According to [STRINGFELLOW ET AL. 90] the Monte-Carlo results for the free energy can be represented by

$$\varepsilon_i = 0.899711\,\Gamma_i - 1.7080\,\Gamma_i^{\frac{1}{3}} - 0.053625\,\Gamma_i^{-\frac{1}{3}} + 0.074823\,\ln(\Gamma_i^3) + 1.29792 \tag{A.8}$$

in the liquid phase $1 < \Gamma_i \leq 178$ and by

$$\varepsilon_i = 0.895929\,\Gamma_i - 4.5\,\ln(\Gamma_i) + 3.94337\,\Gamma_i^{-1} + 1245\,\Gamma_i^{-2} + 1.8856 \tag{A.9}$$

in the solid phase $\Gamma > 178$.

A.3 Electron Gas

For the homogeneous electron gas exchange and correlation effects have to be considered. In the limit of zero temperature the exchange and correlation contribution to the thermodynamic functions are comparatively well understood. Low [WIGNER 34, GELL-MANN & BRUECKNER 56] and high density limits [GELL-MANN & BRUECKNER 56] are known exactly and diffusion Monte-Carlo simulations [CEPERLEY & ALDER 80] can be utilized to fill the gap. Several parametrizations of this data have been suggested, e.g. by [PERDEW 85]

$$\varepsilon_{xc} = \varepsilon_x + \varepsilon_c, \quad \text{with} \quad \varepsilon_x = -\frac{3}{2}\left(\frac{3}{2\pi}\right)^{\frac{2}{3}}\frac{1}{r_s} \quad \text{and} \quad \text{(A.10)}$$

$$\varepsilon_c = \begin{cases} -0.2864\,(1+1.0529\sqrt{r_s}+0.3334\,r_s^{-1} & \text{for} \quad r_s \geq 1 \\ -0.0960+0.0622\ln r_s -0.0232\,r_s +0.0040\,r_s\ln r_s & \text{for} \quad r_s < 1 \end{cases}$$

A.4 Mixture of Hard Spheres

Consider a mixture of hard spheres with densities n_j and diameters d_j. By introducing the the averaged powers of the diameters [FÖRSTER 92]

$$\langle d^k \rangle = \sum_{j=0} \frac{n_j}{n} d_j^k, \qquad k=1,2,3, \qquad n = \sum_{j=0} n_j, \qquad \text{(A.11)}$$

the formula for the interaction contribution to the free energy suggested by [MANSOORI ET AL. 71] can be rewritten in a compact form

$$f_{hc} = n k_B T \left(\frac{X\eta}{(1-\eta)^2} + \frac{3Y\eta}{1-\eta} + (X-1)\ln(1-\eta) \right) \qquad \text{(A.12)}$$

where

$$\eta = \frac{\pi}{6}\frac{N}{V}\langle d^3 \rangle, \qquad X = \frac{\langle d^2 \rangle^3}{\langle d^3 \rangle^2}, \qquad Y = \frac{\langle d^2 \rangle \langle d^1 \rangle}{\langle d^3 \rangle}. \qquad \text{(A.13)}$$

The corresponding expression for a one-component system was first given by [CARNAHAN & STARLING 69]

$$f_{hc} = n k_B T \frac{4\eta - 3\eta^2}{(1-\eta)^2}, \qquad \eta = \frac{4\pi}{3}nr^3 \qquad \text{(A.14)}$$

where $r = d/2$. Note that beyond a certain filling parameter η the hard-sphere system will freeze and equations (A.12) and (A.14) are no longer applicable. For a one component system one finds $\eta_{freeze} \approx 0.494$ [SAUMON ET AL. 89].

B Markoff Automata

This appendix gives the multi-component/multi-dimensional generalization of the stochastic description of reaction-diffusion systems introduced in section 6.2 and describes the method that is used to simulate them. The system is modeled by a discrete MARKOFF process whose dynamic is described by a master equation. The multi-dimensional state of the process is introduced in the first section of this appendix, then the possible transitions are derived from the elementary physical processes. The third part of this appendix summarizes the algorithm used to handle the extensive number of possible transitions whose rates may vary by many orders of magnitudes.

B.1   State of the Process

By discretizing the spatial structure, the d-dimensional volume V is divided into L^d subvolumes or cells of linear size h and volume $\omega = h^d$. The discretization of the concentration is achieved by representing a certain concentration n_v of species v by M_v (artificial) particles. The refinement of this discretization can be controlled by an extensive scaling factor Ω,

$$M_v = \Omega n_v \, . \tag{B.1}$$

If one performs both the limits $h \to 0$ and $\Omega \to \infty$, we regain the continuous and deterministic description of the system, cf. section 6.4. The state of a cell located at position r is defined by the number of artificial particles of each species M_{vr}. The state of the entire system $\mathcal{M}$ is the direct sum of the local states,

$$\mathcal{M} = \{M_{vr}\} \, , \tag{B.2}$$

with v running through the species and r running through the d-dimensional lattice vectors $r = h \cdot (i_1, \ldots, i_d)$ where each of the lattice coordinates $i_1, \ldots, i_d$ takes on the values $1, 2, \ldots, L$.

The dynamics are governed by the multivariate master equation

$$\frac{dp(\mathcal{M}, t)}{dt} = \sum_{\mathcal{M}'} \left[R\left(\mathcal{M}' \rightarrow \mathcal{M}\right) p\left(\mathcal{M}', t\right) - R\left(\mathcal{M} \rightarrow \mathcal{M}'\right) p\left(\mathcal{M}, t\right) \right] \quad \text{(B.3)}$$

that describes the time evolution of the probability $p(\mathcal{M}, t)$ that the system is in state $\mathcal{M}$ at time t. This approach is valid for any reaction-diffusion system. Now one has to give a concrete description of the possible transitions in reacting and diffusing plasma.

B.2   Transition Rates

The state $\mathcal{M}$ may change into $\mathcal{M}'$ by several processes, which can perform local changes in chemical composition or influence next neighbors by diffusion.

Diffusion: If cell r' is a next neighbour to r a diffusion event for a single artificial particle may occur with rate ρ_v. This rate has to be adapted to the diffusion constant D_v, that is assumed to be independent of the density n_v. We approximate this rate by $\rho_v = D_v/h^2$, which results the correct limit for $h \rightarrow 0$. Thus, the jump rate for M_{vr} artificial particles of species v in a single box r is an extensive quantity

$$\rho_v^{(M_{vr})} = M_{vr} \rho_v \ . \quad \text{(B.4)}$$

Therefore we use

$$\begin{aligned} R\left(\mathcal{M} \rightarrow \mathcal{M}'\right) &= M_{vr} \rho_v \\ \text{for} \quad \mathcal{M} &= \{\ldots, M_{vr}, \ldots, M_{vr'}, \ldots\} \\ \text{and} \quad \mathcal{M}' &= \{\ldots, M_{vr} - 1, \ldots, M_{vr'} + 1, \ldots\} \end{aligned} \quad \text{(B.5)}$$

as diffusion-transition rates in the master equation (B.3). As each cell of a d-dimensional grid has $2d$ neighbors the total probability that one of the M_{vr} particles leaves the cell by a diffusive process is $2dDM_{vr}/h^2$.

Chemical Reactions: The chemical reactions are described by local non-linear processes which can be considered as combined generation and decay events. Any chemical reaction γ can be described by the forward and backward coefficients, f_ν^γ and b_ν^γ, for all species X_ν

$$\sum_\nu f_\nu^\gamma X_\nu \rightarrow \sum_\nu b_\nu^\gamma X_\nu \,. \tag{B.6}$$

The chemical reaction γ changes the state from

$$\mathcal{M} = \{\ldots, M_{\nu r}, \ldots\} \qquad \text{to} \qquad \mathcal{M}' = \{\ldots, M_{\nu r} + b_\nu^\gamma - f_\nu^\gamma, \ldots\} \,. \tag{B.7}$$

The rate of the chemical events performed by reaction γ in location r can be given in general by [FRICKE & WENDT 95]

$$R\left(\mathcal{M} \rightarrow \mathcal{M}' \mid \gamma, r\right) = \Omega \lambda^\gamma \prod_\nu \frac{M_{\nu r}!}{(M_{\nu r} - f_\nu^\gamma)!} \Omega^{-f_\nu^\gamma} \,, \tag{B.8}$$

with a reaction specific intensive constant λ^γ. Different reactions may result in the same transition. Therefore, the rate of a transition in r from $\mathcal{M}$ to $\mathcal{M}'$ depends on all reactions performing the same changes

$$R\left(\mathcal{M} \rightarrow \mathcal{M}' \mid r\right) = \sum_\gamma R\left(\mathcal{M} \rightarrow \mathcal{M}' \mid \gamma, r\right) \,. \tag{B.9}$$

B.3 Simulation Algorithm

The algorithm is based on a fact that a discrete MARKOFF process has an exponentially distributed lifetime τ [KARLIN & TAYLOR 75]

$$P\{\tau \mid \mathcal{M}\} = \frac{1}{\tau_{\mathcal{M}}} \exp\left(-\frac{\tau}{\tau_{\mathcal{M}}}\right) \,, \tag{B.10}$$

where $\tau_{\mathcal{M}}$ denotes the mean life time of the state $\mathcal{M}$ given by the inverse sum of the rate of all possible transitions out of $\mathcal{M}$

$$\frac{1}{\tau_{\mathcal{M}}} = \sum_{\mathcal{M}'} R\left(\mathcal{M} \rightarrow \mathcal{M}'\right) \,. \tag{B.11}$$

The probability $P\{\mathcal{M} \rightarrow \mathcal{M}'\}$ to enter the new state $\mathcal{M}'$ coming from state $\mathcal{M}$ is proportional to its contribution to leave $\mathcal{M}$

$$P\{\mathcal{M} \rightarrow \mathcal{M}'\} = \frac{R\left(\mathcal{M} \rightarrow \mathcal{M}'\right)}{\sum_{\mathcal{M}''} R\left(\mathcal{M} \rightarrow \mathcal{M}''\right)} \,. \tag{B.12}$$

This algorithm has been developed several times independently [KARLIN & TAYLOR 75, GILLESPIE 76, GILLESPIE 78, FEISTEL & EBELING 89] and it is widely used for homogeneous systems ($d = 0$).

It is easy to draw a random numbers according to (B.11) from uniformly distributed random numbers $0 \leq u < 1$ by setting $\tau = \tau_{\mathcal{M}} \log(1 - u)$. For a small number of transitions the selection of the state $\mathcal{M}'$ can be implemented easily by summing up (6.30) explicitly. However, for a large number of possible transitions severe complications arise as the extensive sum (6.30) necessary for selecting a new state results in an extensive slow down in computing velocity. For typical lattices with 10^4 to 10^6 cells this would mean a slow down by more than four orders of magnitudes. Another complication is the extreme inhomogeneity of the transition rates caused by their non-linear dependence on the charged-particle densities.

Markoff Automata: The problem of selecting a box and finding an event has been solved in [FRICKE & WENDT 95, BEULE ET AL. 98b]. An alternative approach has been proposed by [GIBSON & BRUCK 00]. The rates belonging to a certain cell are grouped together. Each cell is chosen proportional to the probability that an artificial particle within this cell moves out by diffusion or that a chemical reaction in this cell occurs.

Furthermore, all cells with the same dual order of reaction probabilities are grouped into a logarithmic class. Within a class, the selection of a cell can be performed by a VON NEUMANN rejection [KNUTH 81] with an average efficiency of arrangement of 75%. The number of classes is small compared to the number of cells as a result of the logarithmic construction. Then the transition can be drawn in three steps:

(i) draw a class by summing up the total rates of the classes,

(ii) draw a cell within the chosen class by the rejection algorithm,

(iii) draw an event related to the cells according to the diffusion and reaction probabilities given by the local situation.

This method of drawing an event from an extensive, extremely inhomogeneous set of possibilities, which is called MARKOFF automata, has been successfully applied to other problems [FRICKE & WENDT 95, WENDT ET AL. 95].

It is well known that close to a phase transition the fluctuations dominate the dynamics and that fluctuations can be responsible for new stable states and transitions, cf. [HORSTHEMKE & LEFEVER 84, MALCHOW & SCHIMANSKY-GEIER 85, VAN KAMPEN 92]. When simulating a macroscopic number of real atoms by several 10^2 artificial particles, one has to stay sufficiently far away from the state with zero particles that is unstable in the deterministic sense, but stable in stochastic simulations. Therefore, the extensivety parameter Ω has to be chosen carefully.

C Random Numbers

Random number generators should not be chosen at random. - DONALD KNUTH

Random numbers were needed to perform most simulations in this work and they are generally required in many areas of statistical physics, e.g. stochastic optimization, Monte-Carlo methods or stochastic simulation [EBELING & FEISTEL 82, ALLEN & TILDESLEY 90, SCHNAKENBERG 95]. Random numbers are usually generated by a pseudo random number generator (PRNG). A problem with all PRNG is that pseudo random sequences generated in this way contain weak correlations that may lead to spurious simulation results [FERRENBERG ET AL. 92].

In this appendix a simple and efficient method [BEULE & GROSSE 96] for testing *randomness* is introduced and applied to several widely used PRNGs in order to detect weak correlations in pseudo random sequences and helps to determine which PRNGs are suitable for sampling large discrete spaces.

C.1   Pseudo Random Numbers

A PRNG is an iterative map F of a number (or a set of numbers) x_j onto a new number $x_{j+1} = F(x_j)$. The map is chosen in such a way that for suitable initial values x_0 the sequence $\{x_j\}$ (or parts of it) appear randomly distributed in a certain interval [MARSAGLIA 92]. The pseudo random numbers generated in this way have several desirable features: they are reproducible (e.g. for counter checking results) and produced efficiently without any special equipment[1]. An overview of the many possibilities for choosing the map $F(x_j)$ and the corresponding initial

---

[1]True random numbers can be generated e.g. from thermal electron noise [KNUTH 81].

values is given in [KNUTH 81, MARSAGLIA 92].

Because of (binary) coding in computers the number of different values x_j is limited and PRNGs are periodic. For many PRNGs the length of this period can be determined analytically or at least estimated [MARSAGLIA 92, KNUTH 81]. Besides periodicity the sequences x_j can contain further correlations. For good generators these correlations are quite small and therefore difficult to detect especially because any correlation measure has finite-size effects, that simulate correlations even in truly random finite sequences [HERZEL ET AL. 94, GROSSE 95]. Nevertheless these small correlations - especially if sampling high dimensional or fine structured spaces - may lead to spurious simulation results cf. [FERRENBERG ET AL. 92].

What is a random sequence?   Before proceeding to actual tests for PRNGs a more precise definition of *random* has to be given. Even for infinite sequences it is difficult to define *random* in such a way that on the one hand there are *random* sequences and on the other hand no contradiction to the intuitive understanding of *random* arises cf. [KNUTH 81]. When performing empirical tests of PRNGs one is always restricted to finite sequences. It seems impossible to give a proper definition of *random* for a finite sequence because any sequence of given length has equal probability. However, everybody will agree that the decimal sequence 897932384626 appears to be more *random* than 919191919191, 1234567890123 or 000000000000. A *typical* finite sequence of length N is characterized by the fact that all finite subsequences of length k for any $k \leq \log_\lambda N$ will appear with equal probability $p(k)$. Here λ is the size of the alphabet i.e. the number of different letters in the sequence. There are $M = \lambda^k$ different sequences of length k and therefore $p(k) = 1/M$. Sequences that have the desired distribution for a given k are called k-distributed [KNUTH 81].

An example: Let us consider a PRNG of the widely used class of linear-congruential generators:

$$x_{j+1} = F(x_j) = (a \cdot x_j + c) \bmod m , \qquad j > 0 , \qquad (C.1)$$

where a, c, m and x_j are integer. The quality of a linear-congruential generator (LCG) depends on the proper choice of the parameter a, c, m, x_0. As an instructive example we consider $a = 137$, $c = 187$, and $m = 256$. For any initial value (C.1) will generate equi-distributed (1-distributed) pseudo random numbers of period 256 [KNUTH 81].

In order to randomly sample the fields of a chess board with 8×8 fields one will need 2-distributed coordinates (y_{2j}, y_{2j+1}). These y_j might be generated from the 3 leading bits of x_j i.e. $y_j = x_j >> 5$, where $>>$ denotes the bit-shift operator. It will turn out that the pairs of coordinates (y_{2j}, y_{2j+1}) are not chosen with equal probability. If the coordinates are generated from the last 3 bits of x_j i.e. $y_j = x_j \bmod 8$, one will always select the same field while never reaching any other. The reason for this non-uniform distribution are correlations between x_j and x_{j+1} that cause the pseudo random numbers (x_{2j}, x_{2j+1}) to fill only a fraction[2] of the possible $[0, 255] \times [0, 255]$ space. Therefore the coordinates (y_{2j}, y_{2j+1}) do not properly sample the fields of the chess board. The sequence of y_j coordinates are not 2-distributed and simulations on a 8×8 grid may lead to spurious results.

A better choice of the parameter a, c, m, x_0 (especially larger values for a and m) will improve the situation, but if larger chess boards or higher dimensional spaces are considered the problem will appear again. The effect that was just described is well known for linear-congruential PRNGs [MARSAGLIA 68] and it is even possible to determine the maximal distance between neighboring d-tuple $(x_{j+1}, \ldots, x_{j+d})$ with the help of the semi-empirical spectral-test [KNUTH 81]. For PRNGs that are not of the linear-congruential type one can not perform this test.

C.2   Bit-Decay

In order to test whether a sequence of length N is k-distributed one can determine the probability of any subsequence (serial test) or some subsequences (poker test) of length k and compare it with the expectation value $p(k)$ [KNUTH 81]. Another possibility is to determine the number of sequences that did not appear up to length t [STAUFFER 96, BEULE & GROSSE 96]. In the chess-board example ($\lambda = 8$, $k = 2$) this means to ask: how many fields have not been selected after the coordinates have been generated t times.

For a large number of fields $M = \lambda^d$ the nature of this test can be understood easily by considering the analogy to the radioactive decay. Radioactive nuclei decay randomly, independently of their position and with a fixed rate. Hence the

---

[2]The reason for this is the reconstruction of the attractor of the underlying nonlinear map by means of delay coordinates.

number of remaining (not decayed) nuclei follows the well known exponential decay law. For a random sequence (using the definition given above) all M cells are selected with the same fixed probability $1/M$. For $M \gg 1$ one has a POISSON process and the expectation value $\langle f(t) \rangle$ for the number of cells, that were not selected after t trials decays exponentially:

$$\langle f(t) \rangle \approx M \exp(-t/M) \qquad \text{for} \qquad 1 \ll M = \lambda^k . \qquad (\text{C.2})$$

Not k-distributed sequences lead in general to deviations from the exponential decay law (C.2), cf. [BEULE & GROSSE 96]. Expectation value and standard deviation of $f(t)$ can be given exactly even for finite M, see section C.3.

The information whether a cell has been selected can be stored by a single bit. Therefore the *bit-decay* can be used as an efficient test of the quality of the PRNG generating the sequence. It allows to test whether a given sequence of pseudo random numbers is k-distributed.

C.3 Statistics of the Bit-Decay

The probability distribution $P(f,t,M)$ for occupying $M - f$ of the $M = \lambda^k$ cells in t trials corresponds is given by [FELLER 68]:

$$P(f,t,M) = \binom{M}{f} \cdot \sum_{j=0}^{M-f} (-1)^j \cdot \binom{M-f}{j} \cdot \left(1 - \frac{f+j}{M}\right)^t . \qquad (\text{C.3})$$

In the limit $M \to \infty$ the number of occupied cells is given by a POISSON distribution.

Expectation value: In order to calculate the expectation value for $f(t)$ after t trials consider the binary variable $X_i(t)$ $(i = 1, 2, ..., M)$. Let $X_i(t) = 1$ if the i-th cells is empty after t trials and $X_i(t) = 0$ otherwise. The number of empty cells is given by [FELLER 68]

$$f(t) = \sum_{i=1}^{M} X_i(t) . \qquad (\text{C.4})$$

Therefore the expectation value $\langle f(t) \rangle$ of empty cells after t trials is given by

$$\langle f(t) \rangle = \left\langle \sum_{i=1}^{M} X_i(t) \right\rangle = \sum_{i=1}^{M} \langle X_i(t) \rangle . \qquad (\text{C.5})$$

The probability of selecting a single fixed cell i in a single trial is $p = 1/M$. Therefore the probability that the cell is empty after t trials is: $A(t) \equiv \langle X_i(t) \rangle = (1-p)^t$. This gives the exponential decay law

$$\langle f(t) \rangle = M(1-p)^t = M \exp(-b \cdot t) \quad \text{with} \quad b = \ln(M/(M-1)) . \quad (C.6)$$

Variance: The variance of the number of empty cells $f(t)$ is defined by

$$
\begin{aligned}
\sigma^2(f(t)) &\equiv \langle f(t)^2 \rangle - \langle f(t) \rangle^2 \\
&= \sum_i \langle X_i(t)^2 \rangle + \sum_{i \neq j} \langle X_i(t) \cdot X_j(t) \rangle - \sum_{i,j} \langle X_i(t) \rangle \cdot \langle X_j(t) \rangle \quad (C.7) \\
&= \sum_i (\langle X_i(t)^2 \rangle - \langle X_i(t) \rangle^2 + \sum_{i \neq j} (\langle X_i(t) \cdot X_j(t) \rangle - \langle X_i(t) \rangle \cdot \langle X_j(t) \rangle)) \\
&= \sum_i (\langle X_i(t)^2 \rangle - \langle X_i(t) \rangle \cdot \langle X_j(t) \rangle) + \sum_{i \neq j} cov(X_i(t), X_j(t)) .
\end{aligned}
$$

In each sum every term gives the same contribution because all cells are equal. Therefore one finds for the variance

$$
\begin{aligned}
\sigma^2(f(t)) &= M \cdot (\langle X_1(t)^2 \rangle - \langle X_1(t) \rangle^2) + M \cdot (M-1) \cdot cov(X_1(t), X_2(t)) \\
&= M \cdot (A(t) - A(t)^2) + M \cdot (M-1) \cdot [(1-2p)^t - A(t)^2] \quad (C.8) \\
&= M \cdot A(t) - (M \cdot A(t))^2 + M \cdot (M-1) \cdot (1-2p)^t .
\end{aligned}
$$

The statistics of the *bit-decay* can be related to the finite-size effects of the topological entropy[3], see [BEULE & GROSSE 96] for details.

C.4   Tests

The behavior of the *bit-decay* for a variety of sequences with different correlations has been discussed in [BEULE & GROSSE 96]. Here only the abilities and limitations of two selected PRNGs will be presented:

(i) LCG16807: The linear-congruential PRNG (C.1) with the parameter $a = 16807$, $c = 0$, $m = 2^{31} - 1$, and $x_0 = 1$ defined as the *minimal standard* PRNG in [PRESS ET AL. 92]. The period length of this generator is $2.1 \cdot 10^9$ [MARSAGLIA 92].

---

[3]The limit $q \to 0$ of the RENYI entropies [RENYI 70].

(ii) RAND55: A so called lagged-FIBONACCI PRNG with the iteration

$$x_j = (x_{j-55} - x_{j-24}) \bmod m , \qquad\qquad (C.9)$$

initialized with 55 positive integers and with $m = 2^{32}$ [KNUTH 81]. The length of the period is $7.7 \cdot 10^{25}$ [MARSAGLIA 92].

For the stochastic simulations of 2 and 3-dimensional systems one needs random coordinates that are at least 1-distributed, 2-distributed, and 3-distributed. As all modern PRNGs easily pass tests for 1-distribution *bit-decay* tests for $k = 2$ and $k = 3$ for different λ are performed here. The cell coordinates y_j are again generated from subsequent pseudo random integers x_j. It is important to generate y_j from the leading bits of the x_j as this results in a much better distribution of the y_j, cf. section C.1 and [KNUTH 81]. First the sequences LCG16807 and RAND55 are used to sample a chess board of 256×256 fields, i.e. $\lambda = 256$ and $k = 2$. The number of cells that are empty after t trials is shown in Figure C.1 together with the expectation value $\langle f(t) \rangle$ given by the exponential decay (C.6).

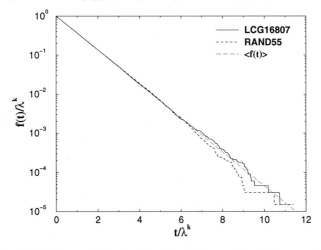

Figure C.1: Bit-decay $f(t)$ for $k = 2$ and $\lambda = 256$ compared with the expected exponential decay (C.6) for the pseudo random sequences RAND55 and LCG16807.

In order to assess whether the observed deviations are are significant one has to compare these deviations with the standard deviation $\sigma(f(t))$ obtained from (C.7).

This comparison is shown in Figure C.2 For the chosen parameter $\lambda = 256$ and $k = 2$ the deviations are within the expected range.   Thus the pseudo random

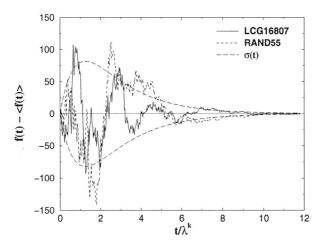

Figure C.2: Bit-decay for $k = 2$ and $\lambda = 256$: deviations of $f(t)$ from the mean $\langle f(t) \rangle$ for the pseudo random sequences RAND55 and LCG16807 compared with the standard deviation $\sigma(f(t))$.

coordinates generated from LCG16807 and RAND55 for the 256×256 grid can be considered 2-distributed according to this test. The situation changes when one considers finer grids (Figure C.3) or higher dimensions (Figure C.4). In these two figures the deviations $f(t) - \langle f(t) \rangle$ are plotted in units of the standard deviation $\sigma(f(t))$.

For LCG16807 one finds significant deviations from the exponential decay up to $+14\,\sigma$ and $-52\,\sigma$. This shows that this generator is not suitable for properly sampling the spaces under consideration. Similar deviations are already found for smaller λ and may lead to spurious results in simulations [BEULE & GROSSE 96]. Lattices where λ is close to a power of 2 are especially prone to this problem. The lagged-FIBONACCI generator RAND55 shows a significantly better performance even for these large high dimensional lattices. Therefore RAND55 was chosen as the standard random number generator for all simulation in this work.

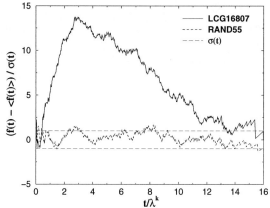

Figure C.3: Bit-decay for $k = 2$ and $\lambda = 3000$: deviations $f(t) - \langle f(t) \rangle$ in units of the standard deviation $\sigma(f(t))$ for RAND55 and LCG16807. Deviations from the exponential decay up to $+14\sigma$ are found for LCG16807. RAND55 only shows deviations in the expected range.

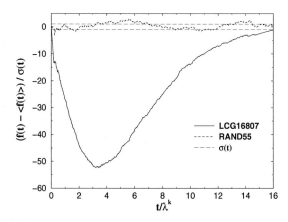

Figure C.4: Bit-decay for $k = 3$ and $\lambda = 300$ deviations $f(t) - \langle f(t) \rangle$ in units of the standard deviation $\sigma(f(t))$ for RAND55 and LCG16807. Deviations from the exponential decay up to -52σ are found for LCG16807. RAND55 only shows deviations in the expected range.

Conclusion: Weak correlation are always present between subsequent pseudo random numbers due to the deterministic character of PRNGs. When sampling fine structured or high dimensional spaces these correlations may lead to suprious results because the sampling becomes non-uniform. The *bit-decay* method allows to detects weak correlations and thus reveals the limitations of popular PRNGs. It can help in selecting proper PRNGs suitable for the planed stochastic simulations.

Acknowledgements

I want to express my gratitude to Prof. Dr. Werner Ebeling for his guidance and for supplying interesting ideas and important know-how during my work. I acknowledge the discussions with Dr. Andreas Förster that helped me to untangle the details of theories and simulations. Furthermore I like to thank Dr. Jens Ortner for useful hints especially on HUGONIOTs and to Dr. Mario Steinberg for discussions and the counter checking of calculations. I would also like to thank Prof. Dr. Lutz Schimansky-Geier for giving helpful hints on stochastic dynamics and Dr. Thomas Fricke for introducing me to efficient stochastic simulation methods.

I want to express my sincere gratitude to Prof. Dr. Ronald Redmer, Prof. Dr. Gerd Röpke, Dr. Stefan Nagel and Dipl.-Phys. Hauke Juranek for the fruitful collaboration on equation of state and to Prof. Dr. Johannes Conrads for the collaboration and discussions on capillary discharges. Special thanks to Prof. Dr. Hanspeter Herzel whose scientific advice inspired my current career. I also thank Dr. Burkhard Militzer for the discussions on HUGONIOTs and for supplying data prior to publication. Furthermore I acknowledge the discussions with Dr. Ivo Große and Dipl.-Phys. Jörg Arndt on randomness and correlations.

This work and the simulations therein were produced using almost exclusively open-source software, therefore I want to thank all the people involved in its development and improvement. Last but not least I want to thank Jessica Holldack and Angelika Schmidt for checking and improving the English syntax and spelling.

Dieter Beule   Berlin, Oktober 2003

Bibliography

[ABE 59] R. Abe, Prog. Theor. Phys. **22** (1959), 213.

[ALASTUEY & PEREZ 96] A. Alastuey and A. Perez, Phys. Rev. E **53** (1996), 5714.

[ALLEN & TILDESLEY 90] M.P. Allen and D.J. Tildesley, *Computer simulation of liquids*, Clarendon Press, Oxford, 1990.

[BARKER & HENDERSON 67] J.A. Barker and D. Henderson, J. Chem. Phys. **47** (1967), 4714.

[BAUMUNG ET AL. 96] K. Baumung et al., Physics of strongly coupled plasmas (W.D. Kreaft and M. Schlanges, eds.), World Scientific, Singapore, 1996, p. 331.

[BAUS & HANSEN 80] M. Baus and J.P. Hansen, Phys. Rep **59** (1980), 1.

[BESPALOV & POLISHCHUK 89] I.M. Bespalov and A.Ya. Polishchuk, Pisma v Zh. Tekhn. Fiz. (USSR) **15** (1989), no. 2, 4.

[BESSON 97] J.M. Besson, Phys. Rev. Lett. **78** (1997), 5026.

[BEULE & GROSSE 96] D. Beule and I. Große, Dynamik, Evolution, Strukturen (J.A. Freund, ed.), Verlag Dr. Köster, Berlin, 1996, p. 161.

[BEULE ET AL. 95] D. Beule, A. Förster, H. Conrads, and W. Ebeling, *Density Effects for Plasmas of Capillary Discharges*, Greifswald, 1995.

[BEULE ET AL. 96] D. Beule, W. Ebeling, and A. Förster, Physica B **228** (1996), 140.

[BEULE ET AL.97] D. Beule, W. Ebeling, and A. Förster, Physica A **241** (1997), 719.

[BEULE ET AL.98a] D. Beule, W. Ebeling, and A. Förster, Strongly Coupled Coulomb Systems (G. Kalman, ed.), Plenum Press, New York, 1998, p. 629.

[BEULE ET AL.98b] D. Beule, A. Förster, and T. Fricke, Z. Phys. Chem. **204** (1998), 1.

[BEULE ET AL.99a] D. Beule, W. Ebeling, A. Förster, H. Juranek, S. Nagel, R. Redmer, and G. Röpke, Phys. Rev. B **59** (1999), 14177.

[BEULE ET AL.99b] D. Beule, W. Ebeling, A. Förster, H. Juranek, R. Redmer, and G. Röpke, Contrib. Plasma Phys. **39** (1999), 21.

[BEULE ET AL.00] D. Beule, W. Ebeling, A. Förster, H. Juranek, R. Redmer, and G. Röpke, J. Phys. IV France, The 1999 Int. Conf. on Strongly Coupled Coulomb Systems in St. Malo (C. Deutsch, B. Jancovici, and M.-M. Gombert, eds.), EDP Sciences, Les Ulis, 2000, pp. Pr5–295.

[BEULE ET AL.01] D. Beule, W. Ebeling, A. Förster, H. Juranek, R. Redmer, and G. Röpke, Phys. Rev. E **63** (2001), 060202(R).

[BIBERMAN ET AL.87] L.M. Biberman, V.S. Vorob'ev, and I.T. Yakubov, *Kinetics of nonequilibrium low-temperature plasmas*, Consultants Bureau, New York, London, 1987.

[BOGEN ET AL.65] P. Bogen, H. Conrads, and D. Rusbüldt, Z. Phys. **186** (1965), 240.

[BOGEN ET AL.68] P. Bogen, H. Conrads, G. Gatti, and W. Kohlhaas, J. Opt. Soc. Am. **58** (1968), 203.

[BOHM & PINES 53] D. Bohm and D. Pines, Phys.Rev. **92** (1953), 609.

[BONITZ & SEMKAT 03] M. Bonitz and D. Semkat (eds.), *Progress in Nonequilibrium Greens Functions II*, World Scientific, Singapore, 2003.

[BONITZ 90] M. Bonitz, *Reaktions- und Diffusions-Prozesse in nichtidealen Plasmen und Entropie für Strukturen im Nichtgleichgewicht*, Ph.D. Thesis, Universität Rostock, 1990.

[BONITZ ET AL. 03] M. Bonitz, D. Semkat, A. Filinov, V. Golubnychyi, D. Kremp, D.O. Gericke, M.S. Murillio, V. Fillinov, V. Fortov, W. Hoyer, and S.W. Koch, J. Phys. A: Math. Gen. **36** (2003), 5921.

[BORNATH ET AL. 98] T. Bornath, M. Schlanges, and R. Prenzel, Phys. Plasmas **5** (1998), 1485.

[BORNATH ET AL. 01] T. Bornath, M. Schlanges, P. Hilse, and D. Kremp, Phys. Rev. E **64** (2001), 26414.

[BREUER ET AL. 94] H.-P. Breuer, W. Huber, and F. Petruccione, Physica D **73** (1994), 259.

[BREUER ET AL. 95] H.-P. Breuer, W. Huber, and F. Petruccione, Europhys. Lett. **30** (1995), 69.

[BRUNET & DERRIDA 01] E. Brunet and B. Derrida, J. Stat. Phys. **103** (2001), 269.

[BUNKER ET AL. 97a] A. Bunker, S. Nagel, R. Redmer, and G. Röpke, Contrib. Plasma Phys. **37** (1997), 115.

[BUNKER ET AL. 97b] A. Bunker, S. Nagel, R. Redmer, and R. Röpke, Phys. Rev. B **56** (1997), 3094.

[CAPITELLI & BRADSLEY 90] M. Capitelli and J.N. Bradsley, *Nonequilibrium Processes in Partially Ionized Gases*, Plenum Press, New York, London, 1990.

[CARNAHAN & STARLING 69] N.F. Carnahan and K.E. Starling, J. Chem. Phys. **51** (1969), 635.

[CEPERLEY & ALDER 80] D.M. Ceperley and B.J. Alder, Phys. Rev. Lett. **45** (1980), 566.

[CHABRIER & POTEKHIN 98] G. Chabrier and A.Y. Potekhin, Phys. Rev. E **58** (1998), 4941.

[CHABRIER & SCHATZMAN 94] G. Chabrier and E. Schatzman, *The equation of state in astrophysics*, Cambridge University Press, Cambridge, 1994.

[CHRISTIAN-DALSGAARD & DÄPPEN 92] J. Christian-Dalsgaard and W. Däppen, Astron. Astrophys. Review **4** (1992), 267.

[CHU & I 94] J.H. Chu and L. I, Phys. Rev. Lett. **72** (1994), 4009.

[COHEN & MURPHY 69] E.G.D. Cohen and T.J. Murphy, Phys. Fluids **12** (1969), 1404.

[COLLINS ET AL. 95] L. Collins, I. Kwon, J. Kress, N. Troullier, and D. Lynch, Phys. Rev. E **52** (1995), 6202.

[COLLINS ET AL. 98] G.W. Collins, L.B. Da Silva, P. Celliers, D.M. Gold, M.E. Foord, R.J. Wallace, A. Ng, S.V. Weber, K.S. Budil, and R. Cauble, Science **281** (1998), 1178.

[CONRADS 67] H. Conrads, Z. Phys. **200** (1967), 444.

[CONRADS 97] H. Conrads, *private communication*, 1997.

[DA SILVA ET AL. 97] L.B. Da Silva, P. Celliers, G.W. Collins, K.S. Budil, N.C. Holmes, T.W. Barbee Jr., B.A. Hammel, J.D. Kilkenny, R.J. Wallace, M. Ross, R. Cauble, A. Ng, and G. Chiu, Phys. Rev. Lett. **78** (1997), 483.

[DEBYE & HÜCKEL 23] P. Debye and E. Hückel, Physikalische Zeitschrift **9** (1923), 185.

[DEWITT 76] H.E. DeWitt, Phys. Rev. A **14** (1976), 1290.

[DHARMA-WARDANA & PERROT 02] M.W.C. Dharma-wardana and F. Perrot, Phys. Rev. B **66** (2002), 014110.

[DIENEMANN ET AL. 80] H. Dienemann, G. Clemens, and W.-D. Kraeft, Ann. Physik (Leipzig) **37** (1980), 444.

[DRAWIN & EWARD 77] H.W. Drawin and F. Eward, Physica C **85** (1977), 333.

[DREIZLER & GROSS 90] R.M. Dreizler and E.K.U. Gross, *Density Functional Theory*, Springer, Berlin, Heidelberg, 1990.

[EBELING & FEISTEL 82] W. Ebeling and R. Feistel, *Physik der Selbstorganisation und Evolution*, Akademie Verlag, Berlin, 1982.

[EBELING & HILBERT 02] W. Ebeling and S. Hilbert, Eur. Phys. J. D **20** (2002), 93.

[EBELING & MILITZER 97] W. Ebeling and B. Militzer, Phys. Lett. A **226** (1997), 298.

[EBELING & NORMMAN 03] W. Ebeling and G. Normman, J. Stat. Phys. **110** (2003), 861.

[EBELING & RICHERT 85a] W. Ebeling and W. Richert, phys. stat. sol. **128** (1985), 467.

[EBELING & RICHERT 85b] W. Ebeling and W. Richert, Phys. Lett. A **108** (1985), 80.

[EBELING & RICHERT 85c] W. Ebeling and W. Richert, Contrib. Plas. Phys. **25** (1985), 431.

[EBELING & SÄNDIG 73] W. Ebeling and R. Sändig, Ann. Phys. (Leipzig) **28** (1973), 289.

[EBELING & SCHAUTZ 97] W. Ebeling and F. Schautz, Phys. Rev. E **56** (1997), 3498.

[EBELING 67a] W. Ebeling, Ann. Phys. **17** (1967), 415.

[EBELING 67b] W. Ebeling, Ann. Phys. **19** (1967), 104.

[EBELING 68a] W. Ebeling, Ann. Phys. **21** (1968), 315.

[EBELING 68b] W. Ebeling, Ann. Phys. **22** (1968), 33,383,392.

[EBELING 68c] W. Ebeling, Ann. Phys. **38** (1968), 378.

[EBELING 68d] W. Ebeling, Physica **40** (1968), 290.

[EBELING 90] W. Ebeling, Contrib. Plasma Phys. **30** (1990), 553.

[EBELING ET AL. 76] W. Ebeling, W.D. Kraeft, and D. Kremp, *Theory of bound states and ionisation equilibrium in plasma and solids*, Ergebnisse der Plasmaphysik und der Gaselektronik, Band 5, Akademie-Verlag, Berlin, 1976.

[EBELING ET AL. 87] W. Ebeling, A. Förster, R. Redmer, T. Rother, and M. Schlanges, XVIII International Conference on Phenomena in Ionized Gases. Invited Lectures (W.T. Williams, ed.), Adam Hilger, Bristol, 1987, p. 40.

[EBELING ET AL. 88] W. Ebeling, A. Förster, W. Richert, and H. Hess, Physica A **150** (1988), 159.

[EBELING ET AL. 89] W. Ebeling, A. Förster, D. Kremp, and M. Schlanges, Phys. A **285** (1989), 149.

[EBELING ET AL. 91] W. Ebeling, A. Förster, V.E. Fortov, V.K. Gryaznov, and A.Ya. Polishchuk, *Thermophysical Properties of Hot Dense Plasmas*, Teubner, Stuttgart, Leipzig, 1991.

[EBELING ET AL. 92] W. Ebeling, A. Förster, and R. Radtke, *Physics of Nonideal Plasmas*, Teubner, Stuttgart, Leipzig, 1992.

[EBELING ET AL. 96a] W. Ebeling, A. Förster, H. Hess, and M.Yu Romanovsky, Plasma Phys. Control. Fusion **38** (1996), A31.

[EBELING ET AL. 96b] W. Ebeling, A. Förster, and V.Yu. Podlipchuk, Phys. Lett. A **218** (1996), 297.

[EBELING ET AL. 03] W. Ebeling, H. Hache, and M. Spahn, *private communication*, 2003.

[ELIEZER ET AL. 86] S. Eliezer, A. Ghatak, and H. Hora, *An introduction to equation of state: theory and application*, Cambridge University Press, Cambridge, 1986.

[FEISTEL & EBELING 89] R. Feistel and W. Ebeling, *Complex Systems and Self-Evolution*, Verlag der Wissenschaften; Kluwer, Berlin; Dordrecht, 1989.

[FELLER 68] W. Feller, *An Introduction to Probability Theory and its Applications*, 3rd ed., John Wiley & Sons, New York, 1968.

[FERRENBERG ET AL. 92] A.M. Ferrenberg, D.P. Landau, and Y.J. Wong, Phys. Rev. Lett. **69** (1992), 3382.

[FILINOV ET AL. 03] V.S. Filinov, M. Bonitz, P. Lavashov, V.E. Fortov W. Ebeling, M. Schlanges, and S.W. Koch, J. Phys. A: Math. Gen. **36** (2003), 6069.

[FÖRSTER 92] A. Förster, *Zusammensetzung, Zustandsgleichung, Phasendiagramm und Ionisationskinetik von stark gekoppelten und partiell ionisierten Plasmen*, Ph.D. Thesis, Humboldt Universität, Berlin, 1992.

[FÖRSTER ET AL.92] A. Förster, T. Kahlbaum, and W. Ebeling, Laser Part. Beams **10** (1992), 253.

[FÖRSTER ET AL.98] A. Förster, D. Beule, H. Conrads, and W. Ebeling, Contrib. Plas. Phys. **38** (1998), 655.

[FORTOV & IAKUBOV 90] V.E. Fortov and I.T. Iakubov, *Physics of Nonideal Plasma*, Hemisphere, New York, 1990.

[FRICKE & WENDT 95] T. Fricke and D. Wendt, Int. J. Mod. Phys. C **6** (1995), 277.

[GALAN & HANSEN 76] S. Galan and J.P. Hansen, Phys. Rev. A **106** (1976), 816.

[GELL-MANN & BRUECKNER 56] M. Gell-Mann and K.A. Brueckner, Phys.Rev. **106** (1956), 364.

[GIBSON & BRUCK 00] M.A. Gibson and J. Bruck, J. Phys. Chem. A **104** (2000), 1876.

[GILLESPIE 76] D.T. Gillespie, J. Comp. Phys. **22** (1976), 403.

[GILLESPIE 78] D.T. Gillespie, J. Comp. Phys. **28** (1978), 435.

[GLEICHAUF 51] P. H. Gleichauf, J. Appl. Phys. **22** (1951), 535,766.

[GRIEM 68] H. Griem, *Plasma Spectroscopy*, McGraw-Hill, New York, 1968.

[GROSSE 95] I. Große, *Statistical Analysis of Biosequences*, Diploma thesis, Humboldt-Universität, 1995.

[HARONSKA ET AL.87] P. Haronska, D. Kremp, and M. Schlanges, Wissenschaftliche Zeitschrift der Wilhelm-Pieck Universität Rostock **36** (1987), 98.

[HENSEL & EDWARDS 96] F. Hensel and P.P. Edwards, Chem. Eur. J. **2** (1996), 1201.

[HERZEL ET AL.94] H. Herzel, A. Schmitt, and W. Ebeling, Chaos, Solitons & Fractals **4** (1994), 97.

[HERZFELD 16] K.F. Herzfeld, Ann. Physik **51** (1916), 261.

[HOHENBERG & KOHN 64] P. Hohenberg and W. Kohn, Phys. Rev. B **136** (1964), 864.

[HOHL ET AL.93] D. Hohl, V. Natoli, D.M. Ceperley, and R.M. Marin, Phys. Rev. Lett. **71** (1993), 541.

[HOLMES ET AL.95] N.C. Holmes, M. Ross, and W.J. Nellis, Phys. Rev. B **52** (1995), 15835.

[HORSTHEMKE & LEFEVER 84] W. Horsthemke and R. Lefever, *Noise-Induced Transitions*, Springer, Berlin, 1984.

[HUBBARD 89] W.B. Hubbard, Simple molecular systems at very high density (A.Polian, L.Loubeyre, and N.Boccara, eds.), Plenum, New York, 1989, p. 203.

[HUBER & HERZBERG 79] K.P. Huber and G. Herzberg, *Molecular Spectra and Molecular Structure Vol IV: Constants of Diatomic Molecules*, Van Nostrand, New York, 1979.

[ICHIMARU 87] S. Ichimaru, Physics Reports **149** (1987), no. 2&3, 93.

[ICHIMARU 94] S. Ichimaru, *Statistical Plasma Physics I+II*, Addison-Wesley, Reading, 1992 and 1994.

[JONES & CEPERLEY 96] M.D. Jones and D.M. Ceperley, Phys. Rev. Lett. **76** (1996), 4572.

[JURANEK & REDMER 00] H. Juranek and R. Redmer, J. Chem. Phys. **112** (2000), 3780.

[JURANEK ET AL.01] H. Juranek, R. Redmer, and W. Stolzmann, Contrib. Plasma Phys. **41** (2001), 131.

[JURANEK ET AL.02] H. Juranek, R. Redmer, and Yaakov Rosenfeld, J. Chem. Phys. **117** (2002), 1768.

[JURANEK ET AL.03] H. Juranek, V. Schwarz, and R. Redmer, J. Phys. A: Math. Gen. **36** (2003), 6181.

[KAGAN ET AL.77] Yu.M. Kagan, V.V. Pushkarev, and A. Kholas, Sov. Phys. JETP **46** (1977), 511.

[KARLIN & TAYLOR 75] S. Karlin and H.M. Taylor, *A First Course in Stochastic Processes and A Second Course in Stochastic Processes*, Academic Press, New York, 1975.

[KERLEY 80] G.I. Kerley, *Los Alamos Scientific Laboratory Report No. LA-4776 (unpublished); J. Chem. Phys. 73, 460 (1980)*, 1980.

[KESSLER ET AL. 98] D.A. Kessler, Z. Ner, and L.M. Sander, Phys. Rev. E **58** (1998), 107.

[KIPPENHAHN & WEIGERT 89] R. Kippenhahn and A. Weigert, *Stellar Structure and Evolution*, Springer, Berlin, Heidelberg, New York, 1989.

[KIRKPATRICK ET AL. 83] S. Kirkpatrick, C.D. Gelatt, and M.P. Vecchi, Science **220** (1983), 671.

[KITAMURA & ICHIMARU 98] H. Kitamura and S. Ichimaru, J. Phys. Soc. Jap. **67** (1998), 950.

[KLAKOW ET AL. 94a] D. Klakow, C. Toepffer, and P.-G. Reinhard, Phys. Lett. A **192** (1994), 55.

[KLAKOW ET AL. 94b] D. Klakow, C. Toepffer, and P.-G. Reinhard, J. Chem. Phys. **101** (1994), 10766.

[KLOSS ET AL. 96] A. Kloss, T. Motzke, R. Grossjohann, and H. Hess, Phys. Rev. E **54** (1996), 5851.

[KNAUP ET AL. 03] M. Knaup, P.G. Reinhard, C. Toepffer, and G. Zwicknagel, J. Phys. A: Math. Gen. **36** (2003), 6165.

[KNUDSON ET AL. 01] M.D. Knudson, D.L. Hanson, J.E. Bailey, C.A. Hall, J.R. Asay, and W.W. Anderson, Phys. Rev. Lett. **87** (2001), 225501.

[KNUDSON ET AL. 03] M.D. Knudson, D.L. Hanson, J.E. Bailey, R.W. Lemke, C.A. Hall, J.R. Asay, and W.W. Anderson, J. Phys. A: Math. Gen. **36** (2003), 6149.

[KNUTH 81] D. Knuth, *The Art of Computer Programming, Vol. 2 Seminumerical Algorithms*, 2nd edition ed., Addision Wesley, Reading, 1981.

[KOBZEV ET AL. 95] G.A. Kobzev, I.T. Iakubov, and M.M. Popovich, *Transport and Optical Properties of Nonideal Plasma*, Plenum Press, New York, London, 1995.

[KOHN & SHAM 65] W. Kohn and L.J. Sham, Phys. Rev. A **140** (1965), 1133.

[KOLMOGOROV ET AL. 37] A.N. Kolmogorov, I.G. Petrovskii, and N.S. Piskunov, Bull. Moscow State University **A 1** (1937), 1.

[KRAEFT & SCHLANGES 96] W.D. Kraeft and M. Schlanges, *Physics of strongly coupled plasmas*, World Scientific, Singapore, 1996.

[KRAEFT ET AL. 86] W.-D. Kraeft, D. Kremp, W. Ebeling, and G. Röpke, *Quantum Statistics of Charged Particle Systems*, Akademie-Verlag, Berlin, 1986.

[KREMP ET AL. 89] D. Kremp, M. Schlanges, T. Bornath, and M. Bonitz, Contrib. Plasma Phys. **29** (1989), 511.

[KUNZE ET AL. 94] H.-J. Kunze, K.N. Koshelev, C. Steden, D. Uskov, and H.T. Wieschebrink, Phys. Lett. A **193** (1994), 183.

[LANDAU & LIFSCHITZ 91a] L.D. Landau and E.M. Lifschitz, *Lehrbuch der theoretischen Physik VI: Hydrodynamik*, Akademie Verlag, Berlin, 1991.

[LANDAU & LIFSCHITZ 91b] L.D. Landau and E.M. Lifschitz, *Lehrbuch der theoretischen Physik X: Kinetik*, Akademie Verlag, Berlin, 1991.

[LARKIN 60] A.I. Larkin, Z. Eksp. Teor. Fiz. **38** (1960), 1896.

[LENOSKY ET AL. 97] T.J. Lenosky, J.D. Kress, and L.A. Collins, Phys. Rev. B **56** (1997), 5164.

[LENOSKY ET AL. 99] T.J. Lenosky, J.D. Kress, L.A. Collins, R. Redmer, and H. Juranek, Phys. Rev. E **60** (1999), 1605.

[LENOSKY ET AL. 00] T.J. Lenosky, S.R. Bickham, J.D. Kress, and L.A. Collins, Phys. Rev. B **61** (2000), 1.

[LEONHARDT & EBELING 93] U. Leonhardt and W. Ebeling, Physica A **192** (1993), 249.

[LINDL ET AL.92] J.D. Lindl, L. McCrory, and E.M. Campbell, Physics Today **45** (1992), no. 9, 32.

[MAGRO ET AL.96] W.R. Magro, D.M. Ceperley, C. Pierleoni, and B. Bernu, Phys. Rev. Lett. **76** (1996), 1240.

[MAI ET AL.96] J. Mai, I.M. Sokolov, and A. Blumen, Phys. Rev. Lett. **77** (1996), 4462.

[MAI ET AL.98] J. Mai, I.M. Sokolov, and A. Blumen, Europhys. Lett. **44** (1998), 7.

[MAI ET AL.00] J. Mai, I.M. Sokolov, and A. Blumen, Phys. Rev. E **62** (2000), 141.

[MALCHOW & SCHIMANSKY-GEIER 85] H. Malchow and L. Schimansky-Geier, *Noise and Diffusion in Bistable Nonequilibrium Systems*, Teubner, Leipzig, 1985.

[MANSOORI ET AL.71] G.A. Mansoori, N.F. Carnahan, K.E. Starling, and T.W. Leland, J. Chem. Phys. **54** (1971), 1523.

[MAO & HEMLEY 94] H.K. Mao and R.J. Hemley, Rev. Mod. Phys. **66** (1994), 671.

[MARLEY & HUBBARD 88] M.S. Marley and W.B. Hubbard, Icarus **73** (1988), 536.

[MARSAGLIA 68] G. Marsaglia, Proceedings of the National Academy of Science **61** (1968), 25.

[MARSAGLIA 92] G. Marsaglia, Proceedings of Symposia in Applied Mathematics Vol. 46 (S. Burr and G. Andrews, eds.), American Mathematical Society, Providence, 1992, p. 73.

[MCQUEEN 91] R.G. McQueen, High-Pressure Equation of State: Theory and Applications (S. Eliezer and R.A. Ricci, eds.), North-Holland, Amsterdam, 1991, p. 101.

[MERMIN 65] N.D. Mermin, Phys. Rev. A **137** (1965), 1441.

[MEYER ET AL. 00] H. Meyer, S. Klose, E. Pasch, and G. Fussmann, Phys. Rev. E **61** (2000), 4347.

[MILITZER & CEPERLY 00] B. Militzer and D.M. Ceperly, Phys. Rev. Lett. **2000** (2000), 1890.

[MILITZER 03] B. Militzer, J. Phys. A: Math. Gen. **36** (2003), 6159.

[MILITZER ET AL. 98] B. Militzer, W. Magro, and D. Ceperley, Strongly Coupled Coulomb Systems (G. Kalman, ed.), Plenum Press, New York, 1998, p. 357.

[MILITZER ET AL. 01] B. Militzer, D.M. Ceperly, J.D. Kress, J.D. Johnson, L.A. Collins, and S Mazevet, Phys. Rev. Lett. **87** (2001), 275502.

[MOORE 93] C.E. Moore, *Tables of Spectra of Hydrogen, Carbon, Nitrogen, and Oxygen Atoms and Ions*, CRC Press, Boca Raton, 1993.

[MORGAN ET AL. 94] C.A. Morgan, H.R. Griem, and R.C. Elton, Phys. Rev. E **49** (1994), 2282.

[MORITA 59] T. Morita, Prog. Theor. Phys. (Japan) **22** (1959), 757.

[MOSTOVYCH ET AL. 00] A.N. Mostovych, Y. Chan, T. Lechecha, A. Schmitt, and J.D. Sethian, Phys. Rev. Lett. **85** (2000), 3870.

[MURRAY 93] J.D. Murray, *Mathematical biology*, Springer, Berlin, Heidelberg, New York, 1993.

[NAGEL ET AL. 98] S. Nagel, R. Redmer, G. Röpke, M. Knaup, and C. Toepffer, Phys. Rev. E **57** (1998), 5572.

[NARAYANA ET AL. 98] C. Narayana, H. Luo, J. Orloff, and A.L. Ruoff, Nature **393** (1998), 46.

[NELLIS & WEIR 97] W.J. Nellis and S.T. Weir, Phys. Rev. Lett. **78** (1997), 5027.

[NELLIS 02] W.J. Nellis, Phys. Rev. Lett. **89** (2002), 165502.

[NELLIS ET AL. 83] W.J. Nellis, A.C. Mitchell, M. van Thiel, G.J. Devine, R.J. Trainor, and N. Brown, J.Chem.Phys. **79** (1983), 1480.

[NELLIS ET AL.92] W.J. Nellis, A.C. Mitchell, P.C. McCandless, D.J. Erskine, and S.T. Weir, Phys. Rev. Lett. **68** (1992), 2937.

[NELLIS ET AL.98] W.J. Nellis, A.A. Louis, and N.W. Ashcroft, Phil. Trans. R. Soc. Lond. A **356** (1998), 119.

[NELLIS ET AL.99] W.J. Nellis, S.T. Weir, and A.C. Mitchell, Phys. Rev. B **59** (1999), 3434.

[NORMAN & STAROSTIN 70] G. Norman and A. Starostin, Teplophys. Vys. Temp. **6** (1970), 410.

[OHDE 97] T. Ohde, *Makroskopische Evolutionsprozesse in dichten reagierenden Plasmen*, Ph.D. Thesis, Universität Rostock, 1997.

[OHDE ET AL.95] T. Ohde, M. Bonitz, T. Bornath, D. Kremp, and M. Schlanges, Phys. Plasma **2** (1995), 3214.

[OHDE ET AL.96] T. Ohde, M. Bonitz, T. Bornath, D. Kremp, and M. Schlanges, Phys. Plasma **3** (1996), 1241.

[OHDE ET AL.97] T. Ohde, M. Bonitz, T. Bornath, and M. Schlanges, Contrib. Plasma Phys. **37** (1997), 229.

[ORTNER ET AL.97] J. Ortner, F. Schautz, and W. Ebeling, Phys. Rev. E **56** (1997), 4665.

[PERDEW 85] J.P. Perdew, Phys. Rev. Lett. **55** (1985), 1665.

[PERROT & DHARMA-WARDANA 84] F. Perrot and M.W.C. Dharma-wardana, Phys. Rev. A **30** (1984), 2619.

[PIERLEONI ET AL.94] C. Pierleoni, D.M. Ceperley, B. Bernu, and W.R. Magro, Phys. Rev. Lett. **73** (1994), 2145.

[PIERLEONI ET AL.96] C. Pierleoni, W.R. Magro, D.M. Ceperley, and B. Bernu, Physics of strongly coupled Plasmas (W.D. Kraeft and M. Schlanges, eds.), Binz 1995, Editors Kraeft, W.D. and Schlanges, M., World Scientific, 1996, p. 11.

[POTEKHIN 96] A.Y. Potekhin, Phys. Plas. **3** (1996), 4156.

144   BIBLIOGRAPHY

[PRESS ET AL.92] W.H. Press, S.A. Teukolsky, W.T. Vetterling, and B.P. Flannery, *Numerical recipes in C*, 2nd ed., Cambridge University Press, Cambridge, 1992.

[RADTKE ET AL.00] R. Radtke, C. Biedermann, T. Fuchs, G. Fussmann, and P. Beiersdorfer, Phys. Rev. E **61** (2000), 1966.

[RECHENBERG 73] I. Rechenberg, *Evolutionsstrategien - Optimierung technischer Systeme nach Prinzipien der biologischen Information*, Friedrich Frommann Verlag, Stuttgart-Bad Cannstatt, 1973.

[REDMER 97] R. Redmer, Physics Reports **282** (1997), no. 2&3, 35.

[REDMER 98a] R. Redmer, *private communication*, 1998.

[REDMER 98b] R. Redmer, Phys. Rev. E **57** (1998), 3678.

[REDMER 99] R. Redmer, Phys. Rev. E **1073** (1999), 59.

[REDMER ET AL.99] R. Redmer, G. Röpke, D. Beule, and W. Ebeling, Contrib. Plasma Phys. **39** (1999), 25.

[REE 88] F.H. Ree, Shock Waves in Condensed Matter - 1987 (S.C. Schmidt and N.C. Holmes, eds.), Elsevier, New York, 1988, p. 125.

[REINHOLZ ET AL.95] H. Reinholz, R. Redmer, and S. Nagel, Phys. Rev. E **52** (1995), 5368.

[RENYI 70] A. Renyi, *Probability Theory*, North-Holland, Amsterdam, 1970.

[RIORDAN ET AL.95] J. Riordan, C.R. Doering, and D. ben Avraham, Phys. Rev. E **75** (1995), 565.

[ROBNIK & KUNDT 83] M. Robnik and W. Kundt, Astro. Astrophys. **120** (1983), 227.

[ROCCA ET AL.94] J.J. Rocca, V. Shlyaptsev, F.G. Tomasel, O.D. Cortázar, D. Hartshorn, and J.L.A. Chilla, Phys. Rev. Lett. **73** (1994), 2192.

[ROGERS & YOUNG 97] F.J. Rogers and D.A. Young, Phys. Rev. E **56** (1997), 5876.

[ROGERS 86] F.J. Rogers, Astrophys. J. **310** (1986), 723.

[RÖPKE 88] G. Röpke, Phys. Rev. A **38** (1988), 3001.

[ROSENFELD 01] Y. Rosenfeld, Contrib. Plasma Phys. **41** (2001), 27.

[ROSS 96] M. Ross, Phys. Rev. B **54** (1996), 9589.

[ROSS 98] M. Ross, Phys. Rev. B **58** (1998), 669.

[ROSS ET AL.83] M. Ross, F.H. Ree, and D.A. Young, J. Chem. Phys. **79** (1983), 1487.

[SAHA 21] M.N. Saha, Z. Physik **6** (1921), 40.

[SAUMON & CHABRIER 91] D. Saumon and G. Chabrier, Phys. Rev. A **44** (1991), 5122.

[SAUMON & CHABRIER 92] D. Saumon and G. Chabrier, Phys. Rev. A **46** (1992), 2084.

[SAUMON ET AL.89] D. Saumon, G. Chabrier, and J.J. Weis, J. Phys. Chem. **90** (1989), 7395.

[SAUMON ET AL.95] D. Saumon, G. Chabrier, and H.M. van Horn, Astro. Phys. J. Supp. **99** (1995), 713.

[SAWADA & FUJIMOTO 95] K. Sawada and T. Fujimoto, J. Appl. Phys. **78** (1995), 2913.

[SCHLANGES & BORNATH 93] M. Schlanges and T. Bornath, Physica A **192** (1993), 262.

[SCHLANGES ET AL.95] M. Schlanges, M. Bonitz, and A. Tschttschjan, Contrib. Plasma Phys. **35** (1995), 109.

[SCHNAKENBERG 95] J. Schnakenberg, *Algorithmen in der Quantentheorie und Statistischen Physik*, Verlag Zimmermann-Neufang, Ulmen, 1995.

[SCHWEFEL 81] H.-P. Schwefel, *Numerical Optimization of Computer Models*, Wiley, New York, 1981.

[SHIN ET AL.94] H.-J. Shin, D.-E. Kim, and T.-N. Lee, Phys. Rev. E **50** (1994), 1376.

[SPITZER 57] L. Spitzer, *Physics of fully ionized Gases*, 2nd ed., Interscience Publisher, New York, 1957.

[STAUFFER 96] D. Stauffer, Computational Physics (K.H. Hoffmann and M. Schreiber, eds.), Springer, Berlin, Heidelberg, 1996, p. 1.

[STOLZMANN & BLÖCKER 96] W. Stolzmann and T. Blöcker, Astron. Astrophys. **314** (1996), 1024.

[STOLZMANN & BLÖCKER 00] W. Stolzmann and T. Blöcker, Astron. Astrophys. **361** (2000), 1152.

[STOLZMANN 96] W. Stolzmann, *Thermodynamics of non-ideal plasmas*, Habilitationsschrift, Universität Kiel, 1996.

[STRINGFELLOW ET AL.90] G.S. Stringfellow, H.E. de Witt, and W.L. Slattery, Phys. Rev. A **41** (1990), 1105.

[TAHIR ET AL.03] N.A. Tahir, H. Juranek, A. Shutov, R. Redmer, A.R. Piriz, M. Temporal, D. Varentsov, S. Udrea, D.H.H. Hoffmann, C. Deutsch, L. Lomononosov, and V.E. Vortov, Phys. Rev. B **67** (2003), 184101.

[TERNOVOI ET AL.99] V.Ya. Ternovoi, A.S. Filimonov, V.E. Fortov, S.V. Kvitov, D.N. Nikolaev, and A.A. Pyalling, Physica B **265** (1999), 6.

[THEOBALD ET AL.96] W. Theobald, R. Häßner, C. Wülker, and R. Sauerbrey, Phys. Rev. Lett. **77** (1996), 298.

[TRIGGER ET AL.03] S.A. Trigger, W. Ebeling, V.S. Filinov, V.E. Fortov, and M. Bonitz, JETP (2003), 465.

[VAN KAMPEN 92] N.G. van Kampen, *Stochastic Processes in Physics and Chemistry*, North Holland, Amsterdam, London, New York, 1992.

[VAN SAARLOOS 89] W. van Saarloos, Phys. Rev. A **39** (1989), 6367.

[WARREN ET AL.01] C.P. Warren, G. Mikus, E. Somfai, and L.M. Sander, Phys. Rev. E **63** (2001), 056103.

[WEIR 98] S.T. Weir, J.Phys.: Condens. Matter **10** (1998), 11147.

[WEIR ET AL.96] S.T. Weir, A.C. Mitchell, and W.J. Nellis, Phys. Rev. Lett. **76** (1996), 1860.

[WENDT ET AL.95] D. Wendt, J. Schnakenberg, and T. Fricke, Z. Phys. B **96** (1995), 541.

[WIGNER 34] E. Wigner, Phys. Rev. **46** (1934), 1002.

[XU & HANSEN 98] H. Xu and J.-P. Hansen, Phys. Rev. E **57** (1998), 211.

[XU & HANSEN 99] H. Xu and J.P. Hansen, Phys. Rev. E **60** (1999), R9.

[ZELDOVICH & RAIZER 66] Y.B. Zeldovich and Y.P. Raizer, *Physics of shock waves and high-temperature hydrodynamic phenomena*, vol. I and II, Academic Press, New York, 1966.

[ZIMMERMANN 88] R. Zimmermann, *Many-particle Theory of Highly Excited Semiconductors*, Teubner, Leipzig, 1988.

[ZINAMON & ROSENFELD 98] Z. Zinamon and Y. Rosenfeld, Phys. Rev. Lett. **81** (1998), 4668.